Costing Adaptation

Preparing for climate change in India

Editors
Anil Markandya • Arabinda Mishra

The Energy and Resources Institute

ISBN 978-81-7993-388-6

[redacted]

Published by
TERI Press
The Energy and Resources Institute
Darbari Seth Block
IHC Complex, Lodhi Road
New Delhi – 110 003
India

Tel. 2468 2100 or 4150 4900
Fax 2468 2144 or 2468 2145
India +91 • Delhi (0)11
E-mail teripress@teri.res.in
Website www.teriin.org

Printed in India

Preface

This book is the result of an ongoing research collaboration agreement between The Energy and Resources Institute (TERI) and the Basque Centre for Climate Change (BC3), published with the aim "*to promote cooperation on climate change and to motivate shared projects and activities in order to achieve that objective*". During the last two years, the partnership has enabled researchers from both sides with varied disciplinary background to work with each other and exchange visits. The experience has been mutually rewarding, and the output has been in the form of a number of joint publications and presentations. This book represents the first major and consolidated outcome of all the effort that has gone into developing this partnership into a meaningful and substantive inter-institutional relationship.

The design of the collaboration between TERI and BC3 follows a phased approach towards identifying activities that are annual and have both research and training components. The subject of this book—estimating adaptation costs for India—was identified as a research priority (the motivation is explained in the introductory chapter) as the first phase activity. The original ambition was to present some figures at Copenhagen in December 2009 during the COP 15 meeting. Data and methodological challenges, however, stretched the study beyond its original timeline. On the other hand, the extra time allowed for a peer review of the methods and findings while they were being finalized.

The research team gratefully acknowledges support by BC3 for the study and associated exchange visits from both sides. The study has been presented, either in part or in its entirety, at different workshops/seminars during the past one year and has benefited from the discussions in such events. We would like to specifically acknowledge the interest and encouragement received from the participants of the Project Seminar at the University of the Basque Country, Bilbao, the Open Seminar at TERI, New Delhi, and the Workshop on Economics of Climate Change Adaptation held at the Madras School of Economics, Chennai.

It would be impossible to name the many individuals who have been consulted by researchers, and the richness of the study owes a good measure to these experts. There are a few TERI colleagues whose contributions appear in this book—Vidyunmala Veldore (for inputs in the agriculture chapter), Priya Dagar and Prasun Gangopadhyay (for the digitization of the maps), and Ridhima Sud (for the table on adaptation funding)—and we are thankful to them for their cooperation.

We would like to acknowledge the assistance received from K Rajagopal, R Sreenivasa Rao, and Savitri Rawat in the administration of the project. Our gratitude also goes to TERI Press for their effort in bringing out this publication. Finally, we are grateful to Dr Leena Srivastava, Executive Director, TERI, for taking a keen interest in the study and providing motivation, and to Dr R K Pachauri, Director-General, TERI, who has, as always, been a source of inspiration and support.

A Markandya
A Mishra

Contents

one

Introduction

Anil Markandya[a] *and Arabinda Mishra*[b]

Introduction

Climate change is expected to have a range of impacts, most of which will be negative for mid- and low-latitude countries like India. These impacts include damage to infrastructure and adverse impacts on human populations, as extreme weather events increase in frequency, more flooding occurs in coastal zones and in some river systems, yields of crops get reduced, water shortages are experienced, and there is increased incidence of water- and vector-borne diseases. All these have been studied in some depth, but there still remain great uncertainties on the magnitude and timing of the impacts. The case for action is strong at two levels: (1) to reduce the likely consequences of climate change by reducing emissions of greenhouse gases (GHGs) and (2) to reduce the consequences of the impacts by adapting to the changes that will inevitably occur. While the former (mitigation) is a global problem, wherein decisions have to be taken globally, the latter (adaptation) is a local problem, wherein decisions have to be taken at the level of the city, river basin or agro-ecological zone.

Societal decisions on adaptation are essentially aimed at "adjustment" to short-term climate variability and long-term

[a] Professor, Ikerbasque, and Director, Basque Centre for Climate Change (BC3), Bilbao, Spain.

[b] Director, Earth Science and Climate Change Division, The Energy and Resources Institute (TERI), and Dean, Faculty of Policy and Planning, TERI University, New Delhi, India.

climate change.[1] Public action[2] on catastrophic climatic phenomena, such as floods, droughts, and cyclones, also involves social choices in allocating resources towards preparedness and emergency response measures. Compared to the reactive action on adaptation, which takes place only after an impact has occurred, planned interventions are based on anticipation and have to deal with uncertainty. For major damages, anticipatory adaptation plays an important role, as complex and long-term measures are needed and should be planned in advance and eventually integrated with reactive measures. For decision-makers committed to planned action on adaptation, cost estimates of suitable measures would be of considerable interest while choosing among the alternatives with large investment needs.[3]

This study looks at what the appropriate adaptation policies and actions are for India and the costs of those policies and actions. Working from the bottom-up, it represents the first attempt at a national review of the measures that are feasible and will be needed in the medium term (from 2030 to 2050) for five key sectors—health, coastal zones, water, agriculture, and forest ecosystems. The literature on adaptation is considerably scarce than that for mitigation, and detailed assessments such as this one are very rare, especially for developing countries. The methodology is still evolving, and a key contribution of this study is the level of disaggregation at which estimates have been generated, the data challenges inherent in disaggregated estimation identified, and operational simplifications carried out to deal with such challenges.

The research on which this book is based was carried out by The Energy and Resources Institute (TERI), in collaboration with

---

[1] The *Intergovernmental Panel on Climate Change (IPCC) Fourth Assessment Report* defines adaptation as the "adjustment in natural or human systems in response to actual or expected climatic stimuli or their effects, which moderates harm or exploits beneficial opportunities" (IPCC 2007).

[2] The Intergovernmental Panel on Climate Change (IPCC) defines public or planned adaptation as the "result of a deliberate policy decision based on an awareness that conditions have changed or are about to change and that action is required to return to, maintain or achieve a desired state", whereas private or autonomous adaptation is defined as "adaptation that does not constitute a conscious response to climate stimuli but is triggered by ecological changes in natural systems and by market or welfare changes in human systems".

[3] Adaptation measures can be put in place to face gradual changes or to face catastrophic events. The former are easier to address and imply a lower cost, while the latter can be difficult to plan for, as such events are difficult to project and have a low probability of occurrence.

the Basque Centre for Climate Change (BC3). The motivation to carry out such a study for India came from the immediate policy context, both at national and international levels. At the national level, India's National Action Plan on Climate Change (NAPCC) is presently guiding the domestic policy response on both mitigation and adaptation in the form of eight National Missions.[4] At least five of these missions relate to sectors that have a strong relevance to the adaptation context. Both agriculture and water are core sectors chosen for this study, and the issues identified under the National Mission on Sustainable Agriculture and the National Mission on Water Resources, respectively, have a bearing on the approach to adaptation adopted here. Similarly, the National Mission on Green India and the National Mission on Himalayan Ecosystem together represent the priority attached to addressing climate change impacts on forests in the country. Health and coastal zones do not figure under the National Action Plan as specific missions, but are addressed in a cross-cutting manner across several sectors. The National Mission on Strategic Knowledge has been set up in response to the weak data and research base in India on impacts, adaptation, and mitigation, and the present study is expected to make a contribution in this regard towards more informed decision-making.

Complementing the National Action Plan at the level of the central government, state-level initiatives that are more recent[5] and focused on local priorities are also being undertaken. These are aimed at mainstreaming climate action into existing policies and programmes. The cross-sector nature of climate change impacts is universally recognized, and efforts are being made towards meaningful institutional arrangements involving inter-departmental linkages and coordination.

---

[4] The eight National Missions are the National Mission on Enhanced Energy Efficiency, the National Mission on Solar Energy, the National Mission on Sustainable Agriculture, the National Mission on Water Resources, the National Mission on Himalayan Ecosystem, the National Mission on Green India, the National Mission on Sustainable Habitat, and the National Mission on Strategic Knowledge.

[5] At the time of this book going to press (September 2010), Orissa became the first Indian state to upload a complete action plan for public review. Among other states, Delhi, Rajasthan, Uttar Pradesh, Maharashtra, West Bengal, and Gujarat have been particularly active in preparing for climate change.

Interestingly, while the National Missions as well as the State Action Plans are coming up with estimates of funding requirements for implementation of the identified measures[6], there is a general lack of transparency on the assumptions and methodology used for such estimates. For adaptation, specifically, a key missing dimension is the comparison across measures for relative cost-effectiveness and administrative suitability.

The international regime on adaptation is slowly but surely emerging as a driver of domestic policy responses to cope with the impacts of climate change. The scale of funding for adaptation in developing and least-developed countries, along with the distribution of the financing burden among rich industrialized nations, continues to be a sensitive issue in international climate negotiations, although there have been significant developments since the Copenhagen meet. More about this is discussed in the concluding chapter of this book.

The structure of the book and its general approach to the topics is simple. Each of the next five chapters focuses on a specific sector to examine feasible adaptation measures to cope with the anticipated risks of climate change, primarily in the short to medium term (meaning next 20 to 40 years). The cost estimates of such measures are largely based on data gathered from ongoing programmes in the country, and the general approach is to apply a bottom-up accounting methodology to get an annualized estimate of the net additional investment required in a sector for building coping capacity by 2030 or 2050. The notion of an adaptation cost curve based on the relative cost-effectiveness of measures is roughly applied to approximate the sequencing plan and to derive the total cost estimate for the time period under consideration. The measures identified for adaptation come from an exhaustive global review of literature, but the choice of options relevant for India is dictated by the country's legal and policy framework. The concluding chapter of the book essentially links up the costing exercise with policy-making

[6] For instance, the National Mission on Green India gives an estimate of total Mission costs amounting to Rs 330 million (approximately $7 million for 10 years; similarly, the National Mission on Water Resources has about Rs 890 million (approximately $19 million) earmarked as the additional funding requirement per year till 2016/17, which marks the end of India's Twelfth Five-year Plan period.

and implementation, specifically focusing on financing issues and the role of institutions.

Sectoral estimates

Human health

In the area of health, we expect an increased incidence of malaria, diarrhoea, and malnutrition by 2030, over a baseline scenario that takes into account the benefits of rapid economic growth in India. The areas where these diseases will occur are identified, and the costs of additional reactive and preventive measures are estimated. The analyses show that the additional annual adaptation costs under different scenarios of development are in the range of $171–546 million in the unmitigated scenario and $141–445 million in a scenario of stabilization at 550 parts per million volume (ppmv) for 2030. Thus, the additional burden on society is considerable, and it falls disproportionately on the vulnerable poorer groups of population having lower adaptive capacity. These figures are little lower than those estimated by the World Bank (2009) study, wherein the costs of adaptation for malaria and diarrhoea in the entire South Asia are estimated at about $850 million/year. Nevertheless, the additional burden is considerable, and most of it will be covered under the public budget, which has not been enough to meet even the current needs of the health sector.

In addition to providing figures at the national level, based on a review of national programmes and strategies being implemented by the Government of India, this study provides state-level estimates that can offer interesting insights to policy-makers for assessing potential adaptation actions at more regional and local levels. Estimates of total costs of adaptation at national level provide useful information for international policy debates on adaptation finance, but a downscaling is essential for state and local governments, given the existing diversity (geographical, demographical, and socio-economic) in different states of India. Future research is nevertheless needed in this line, in order to obtain more robust results based on finer resolution and more comprehensive impact assessment studies.

Our health study also represents the first attempt to distinguish between preventive and reactive measures, at least for the malarial burden. As the results show, the cost of prevention is expected to be much higher than the cost of treatment. Little attention has been paid in previous estimates towards identifying prevention separately from case management, and a single methodology was applied to both. This results in considerable underestimation of the cost, as prevention entails a much larger coverage than just treating the infected population. Another important step related to prevention is the assessment of population expected to be at risk in the future. This was based on some studies available for India in the malarial context, though extensive research is still needed to model these impacts more precisely in different geographical areas.

Forest ecosystems

In the chapter dealing with forest ecosystems, we are able to identify detailed actions to combat the impacts of climate change for the long term only (2075 and later). At that stage, the costs will be substantial (about $1.5–3 billion per annum). Previous studies either ignored this sector or provided estimates based on an ad hoc increase in conservation budgets. In contrast, this study is based on the estimates of actual damages and the costs of making it good where feasible.

In addition to long-term impacts, some non-regret actions are expected to require continued support in the medium term (beyond the current Eleventh Five-year Plan) as a result of climate change. Programmes envisaged under this sector include the National Afforestation Programme (NAP) scheme, the Integrated Forest Protection Scheme (IFPS), the Forest Information Management and Resource Assessment Scheme, and the Forest Infrastructure Development Scheme. Based on these programmes, the estimated annual costs for forest adaptation for 2030 amount to $144 million.

Agriculture

The study on agriculture was more difficult than that on other sectors because the majority of actions are at the farm-level and are very hard to predict. In this study, we have taken two approaches. The first is to estimate likely loss of livelihoods as a result of climate change and assume that as an upper bound, this will have to be addressed through the public budget. Details of the measures for doing this are discussed more elaborately than the previous studies by downscaling the data to the level of agro-climatic zones (ACZ).

Our estimates indicate an annual loss of about $1–1.5 billion and an increase of $1.8–2.2 billion for autonomous adaptation, by 2050. The second approach was to estimate the amount of planned adaptation based on existing programmes and to calculate the additional costs of extending these programmes to areas in which rainfall patterns are expected to change as a result of climate change. Our estimates add up to $54.3–54.8 billion cumulatively, taking into account the impacts up to 2050. Based on the implementation of these programmes over the next 40 years, the annual costs could turn out to be about $1.4 billion per year.

If we compare our estimates with those in other studies, the closest we have is the World Bank 2010 study, which gives figures for South Asia, and from which we have derived an estimate of about $1.3 billion per year, from 2010 to 2050. It may be coincidental, but the estimates from the World Bank study and the present study are remarkably close, although both have many gaps and use different approaches.

In addition to providing an estimate of the costs at the national level, our study has a number of other important results. It is, for example, the first to downscale the information to the ACZ level, although further progress can still be made in the area of autonomous adaptation by using climatic projections at the regional level. While there have been other studies that have attempted to compute farm-level losses, the underlying data for the calculations has been limited. For instance, the Rosenweig and Iglesias study, on the basis of which national results have been obtained, covers only two stations in India. We try to fill this gap by presenting the results in terms of the costs, both for autonomous and planned adaptation. However, to be able to do this, while we have more data observations, we are also dependent on a limited number of studies, impacts, and productivity losses. Integrated assessments have not been carried out and, therefore, a holistic kind of costs that will be incurred, given the losses and the interventions proposed, are still to be arrived at.

Coastal zones

For coastal zones, our study allows for more soft options and is not solely connected to hard engineering-based options used in some other studies. The adaptation options considered are beach nourishment, shelterbelt plantation (which also includes mangrove regeneration), cyclone risk mitigation through cyclone shelters, and other coastal protection measures. The resulting estimates indicate annual costs of about $760 million or about 0.06% of the gross domestic product (GDP) over a period till 2030. The figure is lower than some other figures, possibly because the expected sea level rise is lower (based on the most recent Intergovernmental Panel on Climate Change [IPCC] projections) (IPCC [1990] estimated a cost of about 0.14% of GDP; Asthana [1994] came up with 0.18%; and ADB [1994], with 0.07%; and Deke, Hooss, Kasten *et al.* [2001], with 0.25%). While this study extends previous work by accounting specifically for cyclones and mangrove regeneration, there is still a need to take into account the costs of autonomous adaptation through migration and undertake reactive measures to account for second-order effects of disasters, which will raise the overall cost figure.

Water resources

The costs of adaptation for freshwater systems are derived at the river basin level, wherein estimates have been made for the basins that are likely to face a deficit, and the costs of measures to meet those deficits have been computed. In addition, basins likely to face inland flooding are identified, and costs of measures to address those are estimated. The costing to meet this deficit has been estimated, giving preference for supply-side measures vis-à-vis demand-side ones, but at the same time, not ignoring demand measures. Within the supply-side measures, prioritization of options has been done based on the least cost options and on the extent to which these options can be applied. About 58% of the deficit is met through supply-side measures, and of the rest, about 80% may be met from demand measures. That still leaves a gap of about 8% of the total deficit, or 4.6 billion cubic metres (BCM), which has to be met from measures not considered in this study. This aspect requires further study to be determined in a cost-effective way.

Our cost estimate of $0.6 billion per annum is quite modest compared to, for example, the World Bank estimate of $2.0–5.4 billion per annum. We would attribute this to two factors. First, we have included a number of low-cost demand options to fill the deficit. Second, we have not included measures to fill the whole deficit (as noted above, about 4.6 BCM remains to be accounted for). In addition, we have not accounted for all measures needed to address the inland flooding. Large-scale damages for the strategies that have not been explored include those concerning land use change/floodplains installation and relocation. Disaster preparedness measures, including early warning systems/forecasting systems, communication campaigns, evacuation plans and first aid, immediate shelter provision, and temporary population relocation, have also not been considered. Moreover, interlinking of rivers and the associated costs have not been included. However, this would have been logical, given that certain areas are projected to observe a surplus in run-off compared to a large proportion in the country facing huge deficits. In spite of these limitations, we would argue that our study provides some insights into (1) the required measures at the river basin level and (2) the role of demand-side measures and some low-cost supply-side ones, such as rainwater harvesting and use of treated waste water.

Conclusion

In providing cost estimates of specific adaptation measures, the study provides an economic framework for policy evaluation in India. It is expected to contribute to a better understanding of adaptation to climate change in India, estimating the costs of dealing with it in an effective manner, and keeping a track of the expenditures directed so that poor and vulnerable people are not worse off as a result of the impacts of climate change. Although more needs to be done (and we identify the key gaps), it should help in the negotiations on global funding for adaptation as well as help design cost-efficient and equitable solutions at the state level.

It would be unwise to ignore the methodological inadequacies. Unlike the case of mitigation, where the aim is broadly to reduce emissions of GHGs and where policies can be evaluated on the basis of the cost per tonne of GHG eliminated, such a simple metric is not available in the case of adaptation. At the same time, it is difficult, computationally and data-wise, to carry out a full cost–benefit analysis of each adaptation measure. Hence, this study looks for simplifications in operational methods that can be used to take decisions on adaptation policies at the sectoral level. What is provided here is only the scenario as a whole and its probable solutions. Inevitably, as these methods are applied, detailed questions will arise and have to be dealt with. Nevertheless, the proposals here should provide a basis for estimating costs in the selected sectors—human health, water resources, agriculture, forest ecosystems, and coastal zones.

References

ADB (Asian Development Bank). 1994. **Climate Change in Asia**. [India Country Report, Manila, The Philippines]

Asthana V. 1994. ***Impacts of Greenhouse-induced Sea Level Rise on the Islands and Coasts of India***. New Delhi: Jawaharlal Nehru University

Deke O, Hooss K G, Kasten C, Klepper G, Springer K. 2001. **Economic impact of climate change: simulations with a regionalized climate-economy model**. [Kiel Working Paper No. 1065, Kiel Institute of World Economics]

IPCC (Intergovernmental Panel on Climate Change). 1990. ***Strategies for Adaptation to Sea Level Rise. IPCC Response Strategies Working Group***. The Hague, the Netherlands: Ministry of Transport and Public Works and Ministry of Housing. 122 pp.

IPCC (Intergovernmental Panel on Climate Change). 2007. ***Climate Change 2007: impacts, adaptation and vulnerability. Contribution of Working Group II to the Fourth Assessment Report of the Intergovernmental Panel on Climate Change***, edited by M L Parry, O F Canziani, J P Palutikof, P J van der Linden, C E Hanson. Cambridge, UK: Cambridge University Press. 976 pp.

World Bank. 2009. **The economics of climate change adaptation: final methodology report** [online]. Details available at <http://sisteresources.worldbank.org/INTCC/Resources/MethodologyReport.pdf>, last accessed on 14 August 2009 two

Human health

Aline Chiabai,[a] Sneha Balakrishnan,[b] Gopal Krishna Sarangi,[c] and Shilpa Nischal[d]

Introduction

Climate change is expected to affect human health in different ways, the main impacts involving an increased incidence of vector-, water- and food-borne diseases; malnutrition and undernourishment; injuries and deaths caused by extreme hydro-geological events; and thermal stress due to an increase in the frequency and intensity of heat and cold waves (IPCC 2007; McMichael, Campbell-Lendrum, Kovats, *et al.* 2004).

These impacts, however, will not be equally distributed in the world. Vulnerability to climate change will be affected by a number of factors, such as the level of exposure, sensitivity, and adaptive capacity[1] (IPCC 2001) and is expected to be higher in developing countries. India is among those countries that are already facing enormous challenges due to extreme events such as droughts, land and coastal flooding, and other extreme weather events, in addition to demographic and socio-economic factors such as rapid population growth, poverty, poor health conditions of populace, water scarcity, and inadequate sanitation conditions. Climate change is, therefore, an additional stressor that is expected to increase the burden of diseases,

[a] Research Professor, Basque Centre for Climate Change (BC3), Bilbao, Spain.

[b] Research Associate, Centre for Global Environment Research, Earth Science, and Climate Change Division, The Energy and Resources Institute (TERI), New Delhi, India.

[c] PhD Scholar, TERI University, New Delhi, India.

[d] Associate Fellow, Resource and Development Economics, Resources, Economic Regulation and Global Security, The Energy and Resources Institute (TERI), New Delhi, India.

[1] Adaptive capacity is defined as "the potential or capability of a system to adapt to (to alter, to better suit) climatic stimuli or their effects or impacts" (IPCC 2001).

in terms of increased morbidity and mortality due to diarrhoea, expansion of natural habitats of vectors and pathogens, increased thermal stress due to heatwaves, as well as likely impacts on the existing socio-economic systems.

The health impacts of climate change can be reduced by implementing adaptive policies, programmes, and practices. Adaptation in the health context has emerged as an important issue in the policy arena, especially after the last report of the Intergovernmental Panel on Climate Change (IPCC) in 2007. It is increasingly being recognized that health adaptation should be integrated into development and mitigation policies, especially in developing countries. It is, therefore, crucial to identify the vulnerable population in these countries, the measures that need to be implemented, and the financial resources required for this purpose. Finally, some indicator is needed to understand whether the costs of adaptation are reasonable and the programme is justified on economic grounds. In this context, economists use "cost-effectiveness analysis", which allows priorities to be set up.

This chapter presents a methodological framework, which is used to estimate the additional cost of planned adaptation in the health sector in India for 2030. The approach adopted includes both preventive and reactive measures, under different climate scenarios (with and without mitigation) and also takes into account the influence of socio-economic development, which is expected in a growing economy like India in the near future. Quantification of the costs of health adaptation requires an extensive analysis of long-term data, such as that on the changes in the incidence of diseases that are attributable to climate change, as well as on economic variables, such as unit costs of specific health measures. Such an analysis becomes convoluted in case of a developing country like India, where consistent long-term datasets are lacking and climate change impact and vulnerability studies for health outcomes are rarely available.

The chapter is organized as follows. The section on how climate change is going to affect human health discusses some of the most likely impacts of climate change on human health, along with a summary of study findings specific to India. The next two sections discuss important concepts based on health adaptation literature, along with the findings from some notable studies on cost assessments for health adaptation under different climate scenarios.

"The methodological framework" section presents the methodological framework developed and employed in the analysis, followed by dedicated sections on the three health outcomes assessed. The dedicated sections discuss the general scenario of the health outcome in India, the key preventive and reactive measures currently adopted by national health programmes, data sources for unit costs, and the results. The section "Knowledge gaps and limitations" concludes the findings with some insight into policy implications of the findings. The final section of this chapter discusses the gaps in current knowledge and the limitations inherent to the current cost estimation.

How climate change is going to affect human health

Association between climate factors and health outcomes become evident from the observation that many diseases occur only during certain seasons, or at places with certain type of climate, or from observations of sudden epidemiological eruptions after events of floods or droughts. As seasonal patterns and climate sensitivities of many of the diseases are well known, the question of concern relates to the extent to which changes in climate will influence human health.

Global climate change is expected to affect human health through pathways that vary in intensity, scale (both spatial and temporal), and directness (Figure 1). While the more direct impacts on health include mortality and morbidity due to exposures to thermal extremes such as heatwaves and cold waves; death and injury due to extreme events such as floods, droughts, and cyclones; and increased concentration of certain air pollutants and aeroallergens, the less direct impacts of climate change include changes in the dynamics and distribution of infectious diseases (food-, water-, or vector-borne) and impact on agro-ecological and socio-economic systems. The highest impacts are expected in developing countries, resulting in increased mortality, morbidity, and suffering.

Thermal extremes

The increased frequency and intensity of extreme temperatures may have direct impacts on human health in terms of thermal stresses, such as hypothermia, influenza, cardiovascular and respiratory diseases, heat exhaustion, heat cramps, dehydration, and many

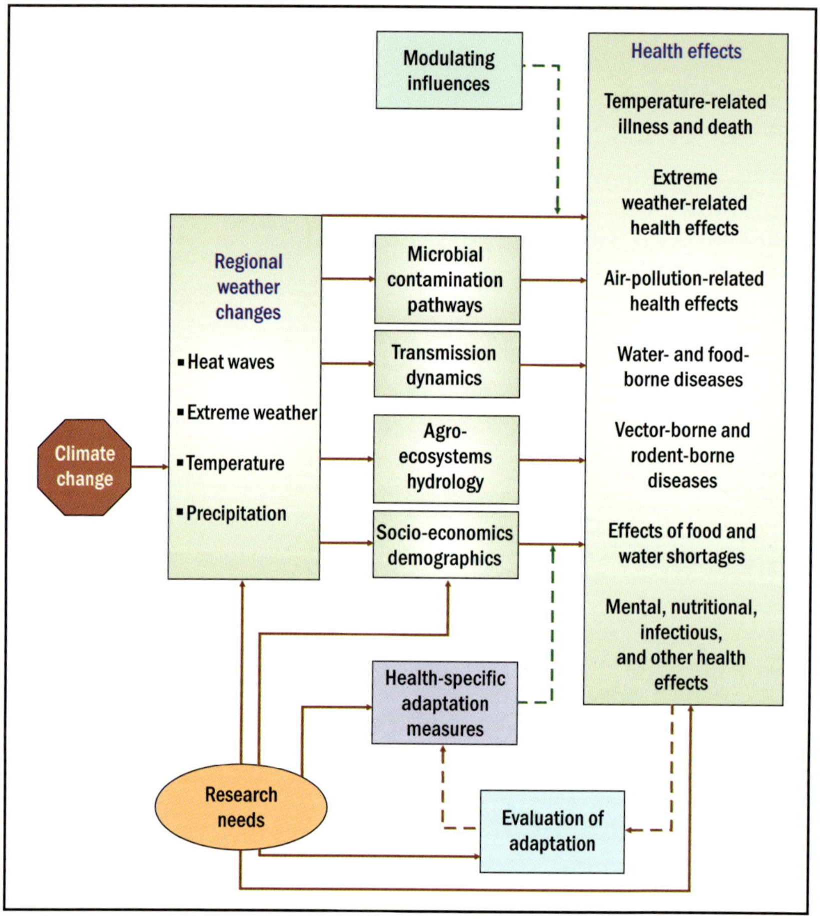

Figure 1 Climate change and health: pathways from driving forces to potential health impacts
Source Adapted from WHO (2003)

others (Faunt, Wilkinson, Aplin, *et al.* 1995; McMichael, Githeko, Carcavallo, *et al.* 2001). High temperatures may also have indirect impacts as they may lead to higher levels of urban air pollution and humidity or exacerbate pre-existing respiratory problems (Gaffen and Ross 1998; Gawith, Downing, and Karacostas 1999; Piver, Ando, Ye, *et al.* 1999). The heatwave of August 2003, which affected many parts of Europe and was linked to several thousand heat-related deaths, can be seen as an indicator of the extremes of summer temperature, which are projected to occur in the near future (Beniston 2003; Kovats and Ebi 2006; Patz, Campbell-Lendrum, Holloway, *et al.* 2005).

Groups at higher risk include the elderly, the children, urban poor, and people with poor health. Urban areas will be particularly affected with the so-called heat island effect that occurs because ambient temperatures in the city become much higher than those in adjoining sub-urban or rural areas, mainly due to concentration of anthropogenic activities and a higher density of concrete structures (Kilbourne 1997; Rooney, McMichael, Kovats, *et al.* 1998). The greatest impact is expected in mid- to high-latitude cities in developing countries, where the existing state of insulation and air conditioning in buildings is poor (for example New Delhi, Mexico City, Jakarta). However, there is a lack of reliable data and long-term statistics in these countries, and only a few studies exist, which project large impacts in the temperate and tropical regions of Asia (Confalonieri, Menne, Akhtar, *et al.* 2007; Epstein, Sohar, and Shapiro 1995; McMichael, Campbell-Lendrum, Kovats, *et al.* 2004).

The increase in heat-related stresses may be compensated by a decrease in cold-related illnesses in the winter season, which as postulated by some studies may lead to an overall positive impact (Donaldson, Kovats, Keatinge, *et al.* 2001; Langford and Bentham 1995; Martens 1997). These effects are nevertheless debatable, as the projected mortality for winter diseases might be underestimated if the effect of influenza is not considered (Armstrong, Mangtani, Fletcher, *et al.* 2004). Further research is needed in this context to identify the impacts under different climatic conditions and socio-economic scenarios. The health benefits, if any, will emerge only in temperate regions, whereas the negative impacts will be much higher worldwide, specifically because of the increased risk of infectious diseases and incidences of malnutrition in developing countries due to increasing trends in temperature.

Air pollutants and aero-allergens

Weather conditions can affect the transportation of air-borne pollutants (reduced atmospheric dispersion and diffusion) and production of certain kinds of pollen. Asthma, allergies, and other acute and chronic respiratory disorders are generally associated with such episodes of air pollution (Haines, McMichael, and Epstein 2000). A study to assess the impact of rising atmospheric CO_2 concentrations on ragweed (*Ambrosia artemisiifolia* L.) pollen production and growth revealed that a doubling of the atmospheric CO_2 concentration stimulated production of ragweed pollen by 61% (Wayne, Foster, Connolly, *et al.* 2002). These results suggest that there may be significant increases in exposure to allergenic pollen under the projected scenarios of global warming. Increased temperature is also expected to increase the frequency and/or the intensity of forest and rangeland fires. The increased particulate exposure due to the resulting smoke and particulate matter may lead to respiratory and cardiovascular conditions (Simkhovich, Kleinman, and Kloner 2008). For ground-level ozone, projections of future morbidity and mortality have been estimated for Europe and North America (Hogrefe, Biswas, Lynn, *et al.* 2004; Stevenson, Johnson, Collins, *et al.* 2000), but there is a lack of information in developing regions despite the high level of pollution in these countries (IPCC 2007).

Extreme events

In recent years, extreme weather events, such as floods and landslides, storms, cyclones, and droughts, have caused considerable damages and loss of life in South Asia (Schultz, Russell, and Espine 2005). The health effects of flooding include deaths and injuries (direct impact), as well as indirect effects, such as mental health problems (anxiety and depression) and increased water-borne diseases (for example, diarrhoea, cholera, and typhoid) due to water contamination (Ahern, Kovats, Wilkinson, *et al.* 2005; Greenough, McGeehin, Bernard, *et al.* 2001). Unsafe sanitation conditions and decreased water accessibility further increase the chances of transmission of water-borne infections. Crowding, common in populations displaced by climatic disasters like floods, can further facilitate the transmission of communicable diseases (Watson, Gayer, and Connolly 2007). Extreme events of weather can also severely disrupt the public health care facilities as well as emergency support services. These can temporarily lead to suspension of disease control programmes due to the inaccessibility of the affected region or due to economic crises arising after an event of a climate disaster or epidemic. Of particular concern are the impacts of sea-level rise, especially in the coastal and low-lying areas, where population displacement and mass migration may impact the socio-economic conditions of both the affected region and the region to which migration takes place (McMichael, Campbell-Lendrum, Corvalan, *et al.* 2003). Drought events may increase malnutrition and undernutrition due to loss in agricultural production and lead to an increase in infectious and respiratory diseases (Menne and Bertollini 2000). It is estimated that by 2050, in Asia, 1 billion people would have limited access to drinking water (Christensen, Hewitson, Busuioc, *et al.* 2007). Droughts can also increase the transmission of some mosquito-borne diseases,[2] whereas stagnation of water in small rivers and drainage canals can contaminate water.

Infectious diseases

Other indirect impacts of changes in weather patterns include variation in the geographical and seasonal distribution of vector-borne diseases[3] (malaria, dengue, Lyme disease, tick-borne encephalitis). The risk of malaria might increase in some regions in central Asia and India (Bhattacharya, Sharma, Dhiman, *et al.* 2006), whereas the transmission could decrease because of reduced rainfall in other areas (Central America and Amazon) (van Lieshout, Kovats, Livermore, *et al.* 2004). Some regions are expected to have a longer transmission season, with important implications for vector control programmes. Populations at the margins of the current distribution of diseases might be particularly affected because such populations have not evolved natural coping mechanisms unlike populations in endemic areas.

However, there is still considerable uncertainty regarding the magnitude of the overall impact of malaria. Some models project a net increase in the number of people exposed to malaria and dengue transmission as a result of climate change (Martens, Jetten, Rotmans, *et al.* 1995), whereas other models suggest only little change in malaria distributions (McMichael, Campbell-Lendrum, Kovats, *et al.* 2004). On the one hand, this uncertainty is due to the complex dynamics of the disease, and on the other, it is due to the ever changing socio-economic conditions. Dengue is expected

[2] The reduced mosquito activity during the drought will cause an increase in non-immune population.

[3] Vector-borne diseases are infections that are transmitted by mosquitoes, ticks, bugs, and sand flies. They are particularly sensitive to climatic factors.

to increase in the urban areas in some parts of Australia and New Zealand (Woodruff, Hales, Butler, *et al.* 2005) as a result of high rainfall and high temperature conditions. Lyme disease could expand to the northern areas in North America (Ogden, Maarouf, Barker, *et al.* 2006), whereas, in Europe, the effect of climate change on Lyme disease and tick-borne encephalitis is still unclear (Rogers and Randolph 2006).

Finally, higher temperatures can lead to increased exposure to water- and food-borne diseases. This may be attributed to the direct effect of temperature on the growth and multiplication of disease-causing organisms in the environment (Checkley, Epstein, Gilman, *et al.* 2000; Singh, Hales, de Wet, *et al.* 2001). According to some studies, increasing sea surface temperature can also directly influence the viability of the enteric pathogen, *Vibrio cholerae*, by creating favourable conditions for the growth of certain blue-green algae, zooplankton, and copepods, which act as storehouse for the pathogen (Colwell 1996; Lipp, Huq, and Colwell 2002). This increases the incidence of gastrointestinal poisoning due to the ingestion of contaminated water or indirectly due to the consumption of contaminated fish that may accumulate these toxins. Although observation shows that the incidence of enteric diseases generally rises with rising temperatures, the reasons for this may vary; for example, in some cases, seasonal lack of food may lead to seasonal malnutrition, and hence, increase the susceptibility to infectious food-borne disease, whereas, in others, temperature and rainfall may have direct effects on the growth and multiplication of pathogens (Hall, D'Souza, and Kirk 2002).

Malnutrition

It is projected in a range of studies that climate change may lead to increase in crop yield at higher latitudes, but at the same time, could lead to a decrease in yield in the tropics and subtropics, aggravating hunger and poverty in these areas. The risk of increasing extreme events of temperature and rainfall is bound to have negative consequences. In some areas, warmer temperatures may offer opportunities to cultivate new crops, but in other areas, rising temperatures may make it more difficult to grow traditional and culturally acceptable staple crops. Floods are a particularly challenging problem, because, apart from damaging crops, they also create favourable breeding conditions for agricultural pests, such as insects, blights, and fungi, and cause soil erosion. Another consequence of climate change, as quoted in many studies, is the "carbon fertilization" effect resulting from higher ambient CO_2 levels. It is postulated that this increase in productivity may offset some of the losses due to other impacts, but more research is needed in this regard (Parry 2007) as the beneficial effects of elevated CO_2 must be considered in the light of the accompanying changes in air temperature, moisture availability, survival and distribution of pest populations, and frequency and intensity of inter- and intra-seasonal droughts and floods, along with soil organic matter transformations, soil erosion, decline in arable areas due to the submergence of coastal lands, and so on (Mall, Singh, Gupta, *et al.* 2006). There have been several studies examining the relationship between climate change, agriculture, and hunger. Tubiello and Fischer (2007) also looked at the relationship between climate, agriculture, and malnutrition. They found that due to climate change, an additional 70–90 million people are likely to be at risk of hunger, with most of them residing in Africa. Liu, Steffen, van Wesenbeeck, *et al.* (2008) carried out an impact assessment study for sub-Saharan Africa to identify hunger "hotspots". They found that in most areas, detrimental effects to one crop, like wheat, were somewhat offset by improved conditions for another, like cassava. However, there are pockets where all crops grown would be adversely affected, increasing the risk of hunger. As per the International Food Policy Research Institute (Nelson, Rosegrant, Koo, *et al.* 2009) in South Asia, a climate scenario may result in a 14% decline in rice production relative to a no climate change scenario, a 44%–49% decline in wheat production, and a 9%–19% reduction in maize production by 2050. However, Parry, Rosenzweig, Iglesiasc, *et al.* (2004), who studied the effects of climate on agriculture and malnutrition, found that the choice of future scenario greatly influences the fact whether climate change will increase malnutrition. Under the worst case scenarios with rapid population growth and no mitigation, the number of persons at risk of hunger will double. Under the scenarios of lower population growth and some mitigation, the number of persons suffering from malnutrition may remain constant.

In general, developing countries are more vulnerable, with lower adaptive capacity due to the dismal condition of existing health care systems, poor health of the population, and inadequate financial, social, and technological resources. Current projections indicate an increase in health impacts, which can be attributed to climate change. On the one hand, climate change is threatening the possibility of attaining the United Nations Millennium Development Goals (MDGs) by 2015 (UN 2008), while on the other, an emerging economy such as India is expected to benefit from the economic growth that will increase social welfare and technological development in the next decades. It is, therefore, crucial to understand how climate change is likely to impact development priorities in poor countries so as to respond with appropriate and cost-effective measures for adaptation.

Impacts expected in India

According to some observations of past and present climatic trends, there has been a marked increase in average surface temperatures in India during the past 100 years, along with regional increasing or decreasing trends in rainfall (GoI 2004).

De and Mukhopadhyay (1998) reported on the severe heatwave that occurred in May 1998, when the condition started in the north-western part of the country and gradually progressed to the south and south-eastern coastal states, causing a large number of heat-related causalities in these regions. Chaudhury, Gre, and Ray (2000), who assessed the impact of recent heatwaves in India, found that the maximum impact of heatwaves was felt in regions where the normal maximum temperature itself is more than 40°C (mostly the north-western region). Although the maximum frequency of heatwaves over most of the states was in the month of May, the number of heatwave condition over Bihar, Rajasthan, and Uttar Pradesh was more during the month of June compared with other months. They also reported that casualties due to heat stress were more in years succeeding an El Niño year over India. Takahashi, Honda, and Emori (2007) also projected an increase in mortality due to thermal stresses in India. Apart from severe direct implications on health, prolonged heat may also lead to water scarcity and loss of soil moisture, which can adversely affect agricultural processes and production, especially if there is a delay in the annual monsoon.

Hajat, Armstrong, Gouveia, *et al.* (2005), who looked at heat-related mortality in the Indian city of New Delhi, revealed an increase in daily all-cause mortality, with the same day and previous day temperatures greater than a threshold of 20°C. McMichael, Wilkinson, Sari Kovats, *et al.* (2008) examined heat-related mortality in urban populations of 12 cities worldwide and found that in New Delhi, the death rates were low during the coolest periods of the year and highest just after monsoon. Heat-related fluctuation in mortality was found to be high, with a positive association between temperature and mortality throughout the temperature range of the city. The percentage increase in mortality (all-cause) was 3.94 for each degree Celsius increase in temperature above the city's heat threshold—one of the highest amongst the 12 cities under analysis. A separate analysis of cardio-respiratory and non-cardio-respiratory diseases showed that the percentage increase in mortality for each degree Celsius increase above "heat threshold" was 3.94 and 4.30, respectively. The increase in respiratory diseases can be attributed to air pollutant–temperature interactions that are augmented by higher temperatures. Thus, with temperatures rising and heatwave duration and intensity increasing, a manifold increase in heat-related disorders can be expected.

There has also been a notable increase in extreme rainfall events during summer monsoon in recent decades in the Indian subcontinent, as reflected by incidences of serious and recurrent floods in Bangladesh, Nepal, and north-eastern states of India during 2002, 2003, and 2004; a record rainfall of 944 mm in the city of Mumbai in 2005, which led to a loss of over 1000 lives; and floods in Surat, Barmer, and Srinagar during summer monsoon season of 2006 (Cruz, Harasawa, Lal, *et al.* 2007). Also, though the frequency of monsoon depressions and cyclone formation in Bay of Bengal and Arabian Sea has been on the decline since 1970, the intensity is increasing, causing severe floods and large-scale damages to life and property (Lal 2001, Raghavan and Rajesh 2003).

As temperature and humidity are key determinants in the transmission of many infectious diseases, such as malaria and diarrhoea, changes in the patterns of temperature, rainfall, and extreme events can lead to an impact on the incidence of these diseases. Bhattacharya, Sharma, Dhiman, *et al.* (2006) projected, in their study, malaria transmission to new geographical regions in India. According to this study, malaria transmission is projected to move from central regions towards south-western and northern regions by 2050. Apart from the spatial shifts, some temporal shifts in the transmission season were also reported. Estimates of malaria impacts at global and regional levels are available in van Lieshout, Kovats, Livermore, *et al.* (2004), who projected the additional population at risk of malaria in different climate change scenarios. Akhtar, Dutt, and Wadhwa (2002), who studied the shifts in the patterns of malarial incidence in urban areas of India, found that despite all efforts made by the state to control malaria, the disease was still widespread in the country, with urban areas becoming targets for the vector, as the urban setting favoured the development of the vector within slum areas, construction sites and green belts acting as habitats for the vector. Thus, any change in the climatic variables towards a more favourable condition for malarial transmission would lead to an increased burden on the urban health infrastructure.

When effects of seasonality and other temporal patterns on the occurrence of rotavirus diarrhoea were studied among hospitalized patients at Pune, India (Purohit, Kelkar, and Simha 1998), the statistical model suggested a strong influence of climatic changes on the incidence of the disease, with daily minimum temperature acting as the principal factor. Flood-related increases in diarrhoeal disease have been reported in India (Mondal, Biswas, and Manna 2001; Sur, Dutta, Nair, *et al.* 2000). Hijioka, Takahashi, Matsuoka, *et al.* (2002) provide estimates of the climate change impact on diarrhoeal diseases for 14 regions across the world and show that diarrhoea tends to increase with temperature.

In India, substantial research has also been conducted for understanding the impact that climatic change might have on different crop yields and overall food security of the country (Kumar 2007; Mall, Singh, Gupta, *et al.* 2006). Most of the simulation studies have shown a decrease in the duration and yield of crops as temperature increased in different parts of India (Aggarwal and Mall 2002). These reductions were, however, generally offset by the increase in CO_2; the magnitude of this response varies with crop, region, and the climate change scenario chosen. Under an optimistic scenario,[4] the effect on the rice yield remains positive (5%–20%) up

---

[4] The optimistic scenario includes a significant increase in CO_2 and a negligible increase in temperature, whereas the pessimistic scenario reflects low increase in CO_2 and a high increase in temperature.

to 2070. In comparison, the effect on wheat may either be positive (up to 25%) or negative (up to 30%), depending on the magnitude of change in CO_2 and temperature. This impact assessment analyses when extended to different climate change scenarios of 2100 show that irrigated rice yields register a small gain, irrespective of the scenario at all regions in India. Wheat yields in central India are likely to suffer by up to 2% in the pessimistic scenario, but there is also a possibility of a beneficial effect (yield increase up to 6%) if the global change is optimistic (GoI 2004). Because most of the cultivated land in India is rain-fed, and agricultural success heavily depends on the annual monsoons, agriculture performance and well-being of the Indian farmer are highly climate-sensitive.

As India is one of the most populated countries in the world, population growth becomes an important additional aspect that needs to be considered. The highest population growth in the next 50 years is expected in South Asia, with Indian population projected to increase by 570 million (IPCC 2007). This is likely to increase the pressure on natural and socio-economic resources (in particular the access to water, food, health care system), which, in turn, will exacerbate the negative impacts of climate change on population health.

Adaptation strategies in the health sector

To reduce the undesirable health impacts of climate change, various adaptation measures or strategies can be implemented. Adaptation in the context of health includes interventions that may be defined as "actions that involve making changes to the natural or human environment or to human behaviour that have beneficial impacts (or prevent adverse impacts) on the health of humans" (Hutton 2000). Besides environmental health interventions, adaptation programmes for health also include adopting typical public health measures like curative interventions.

Adaptation can be defined according to the purpose (autonomous and planned), the timing (preventive and reactive), the temporal scope (short and long term), the spatial scope (localized and spread), the form (legal, technical, advisory, and behavioural), the function (structural and non-structural), and the valuation of performance (effectiveness-feasibility) (EEA 2007). Autonomous (or private) adaptation results from "changes that economic agents make when confronted with climate change" (Aaheim and Aasen 2008). Planned adaptation, on the other hand, results from deliberate policy decisions by public actors, which is necessary to achieve a social level of adaptation and welfare. With respect to the timing of intervention, preventive adaptation consists of measures put in place before the impact occurs, with the purpose of reducing or avoiding the risk. The interventions in this context entail decisions over a longer time period to cope with impacts expected to occur in the long run. Preventive adaptation is, therefore, important for reducing the impact and requires plans or programmes designed by the government. Reactive actions, on the other hand, are put in place after the impact occurs, to minimize the expected damage. These actions provide an immediate response to climate change.

In the public health context, adaptation measures can be classified into three categories (McMichael, Campbell-Lendrum, Kovats, *et al.* 2004; McMichael and Kovats 2000). The first is primary prevention, which includes measures that are set up before the impact occurs to reduce the exposure (for example, warning systems, distribution of impregnated bed nets, flood protection structures). The second group includes secondary prevention, wherein the measures are implemented after the impact but before the episode, and they can involve non-structural actions, such as the monitoring of the disease and its early detection (for example, check up and testing), or structural actions such as infrastructural improvements in the public health care systems. The third, tertiary prevention consists primarily of curative treatments aimed at avoiding deaths and reducing negative consequences of the disease for an early recovery of the patient (for a discussion, see Markandya and Chiabai 2009).

Adaptation can include legislative, technical, educational, and behavioural measures. The first consists of laws, regulations, and directives meant to establish some guidelines, involving, for example, urban planning regulation, water quality and supply regulation, hygiene regulation, and so on. Technical adaptation includes structural or non-structural interventions, depending on the type of change they involve, infrastructural or not (examples of the former include water and sanitation system construction and flood protection structures, whereas the latter includes vaccination, impregnated bed nets, and epidemic forecasting). Educational measures consist of watch warning systems, educational and awareness creation campaigns, food safety, and so on, put in place with the cooperation of the public health bodies. Finally, behavioural adaptation involves measures through which individuals autonomously adapt to the changes (for example, appropriate clothing during heat waves, use of air conditioning in houses, water storage practices, hygiene habits, and so on).

Table 1 reports different types of adaptation measures, according to the features discussed, for the following health outcomes: thermal stresses, extreme weather events, vector-borne diseases, and water- and food-borne diseases.

Table 1 Health adaptation measures to combat climate change

Adaptation	Health impacts				
	Thermal stresses	Extreme weather events	Vector-borne diseases	Water-borne diseases	Food-borne diseases
Legislative (Anticipatory)	Building guidelines	▪ Building guidelines ▪ Economic incentives for building protective structures ▪ Urban planning regulation ▪ Forced migration	▪ Malarial treatment guidelines ▪ Fines against general public for not following vector-control practices, for example, letting water stagnate nearby	▪ Watershed protection laws ▪ Water quality and water supply regulation	▪ Food sanitation and hygiene regulation
Technical (Anticipatory)	▪ Urban planning (green islands, fountains, and green roofs) ▪ Thermal building insulation and air condi-tioning	▪ Urban planning (flood-resistant) ▪ Flood protection elevation ▪ Flood protection structures (dams, dykes, walls	▪ Vector control ▪ Vaccination, impreg-nated bed nets ▪ Surveillance, prevention, and control programmes ▪ Epidemic forecasting	▪ Water treatment and distribution ▪ Monitoring water sources ▪ Regulated piped water in houses – Improved sanitation (latrines)	▪ Refri-geration ▪ Chlorina-tion of drinking water ▪ Pasteuri-zation of milk ▪ Sanitary slaughter and proce-ssing

Contd...

Table 1 *Contd...*

Adaptation	Health impacts				
	Thermal stresses	Extreme weather events	Vector-borne diseases	Water-borne diseases	Food-borne diseases
		and raised banks, and pump stations) ▪ Reforestation		▪ Household sewer connection ▪ Surveillance, prevention and control programmes	of meat, poultry, and seafood
Reactive	▪ Financial and domiciliary assistance services, 'telecare' systems, accompaniment and transport to emergency medical services ▪ Emergency plans (hospital and primary care)	▪ Pre-disaster recovery plans ▪ First aid and emergency plans ▪ Temporary evacuation	▪ Hospital and primary care ▪ Outreach doctors	▪ Hospital and primary care ▪ Outreach doctors	▪ Food-borne disease surveillance ▪ Hospital and primary care ▪ Outreach doctors
Educational (Anticipatory)	▪ Heat watch warning systems ▪ Educational campaign	▪ Real-time forecasting ▪ Early warning systems ▪ Educational campaign	▪ Awareness campaign for vector control ▪ Awareness related to prophylactic measures like use of mosquito repellents in homes	▪ Health educational campaigns ▪ Boil water alerts	▪ Food safety education

Contd...

Table 1 *Contd...*

Adaptation	Health impacts				
	Thermal stresses	Extreme weather events	Vector-borne diseases	Water-borne diseases	Food-borne diseases
Cultural and behaviour (Anticipa-tory)	Clothing, drinking fluids, visiting places with air conditioning, and green areas	Use of storm shelters	Water storage practices	▪ Washing hands and personal hygiene ▪ Use of pit latrines	▪ Avoid risky food (such as runny eggs and raw shellfish) ▪ Separa-ting cooked and raw food ▪ Washing hands, cutting boards, and conta-minated surfaces

Source Adapted from Alberini and Chiabai (2007); IPCC (2001).

Costing adaptation: findings from the literature

Despite the crucial role of adaptation in the health context, there are very few studies available in the literature on costing health adaptation. This section reports two main sets of studies identified from the literature review. The first reports on the costs of taking adaptation measures to reduce or to cope with additional health impacts due to climate change. The second comprises the estimates of the costs of health interventions, focusing on climate-related diseases (malnutrition, diarrhoea, malaria, pneumonia, and diseases in newborns), though outside the climate change context. Because the types of measures are assumed to be similar in both contexts, costs of specific health interventions can be used as indicators of adaptation costs. Thus, the second group of studies can provide interesting insights while assessing adaptation costs.

Some estimates from global and regional adaptation studies

Only few studies are available, which focus directly on the cost of reducing the additional health impacts of climate change. The study of United Nations Framework Convention on Climate Change (UNFCCC) provides some initial estimates in this context (UNFCCC 2007). The estimates include specific interventions for the treatment of additional cases of malaria, diarrhoea, and malnutrition expected to occur between 2000 and 2030, due to climate change. These were estimated under three exposure scenarios: unmitigated emission trends; stabilization at 750 parts per million (ppm) CO_2 equivalent by 2210; and stabilization at 550 ppm CO_2 equivalent by 2170. Estimates are provided for different World Health Organization (WHO) regions, including Africa, North and South America, South East Asia, Europe, Eastern Mediterranean countries, and Western Pacific countries. One of the limitations of the study is that it assumes that the current annual number of cases of disease would remain constant till 2030. Furthermore, the estimated figures do not include the full range of costs; for example, infrastructure, equipment and health-care personnel costs, training costs, and maintenance costs are excluded. Adaptation measures for diarrhoea include immunization and improvement in water supply and sanitation. In this respect, it might be controversial to include the cost of improved water and sanitation structures, as by 2030, some of the currently defined developing countries can be reasonably expected to have a higher level of economic development and gross domestic product (GDP), and would probably have already established water and sanitation systems. For malaria, the identified measures were impregnated bed nets, indoor spraying, case management, and preventive treatment in pregnancy. Finally, in the case of malnutrition, the costs included are for nutritional programmes and growth monitoring. Total costs in the unmitigated scenario are estimated to be in the range of \$3100–8800 million (2000 price) for malaria, \$2731–9010 million for diarrhoea, and \$62–166 million for malnutrition, worldwide.

The recent "Economics of Adaptation to Climate Change" study (World Bank 2009), which estimates the cost of adapting to an approximately 2°C warmer world by 2050, computes health adaptation costs, again on a disease-specific basis for malaria and diarrhoea. The relative risk due to climate change is applied to the projected baseline incidence to determine the additional number of cases attributable to climate change and, for malaria, to determine the number of disability adjusted life years (DALYs) attributable to climate change. However, unlike earlier estimates of the global cost of adaptation, this study does not keep the baseline incidence of disease fixed at current levels. It incorporates a future baseline disease burden (based on WHO projections for 2030), which implies a reduction in the incidence rates that may be primarily attributed to developments that the health sector would undergo in the future. As per the findings, the average annual adaptation costs in the health sector for diarrhoea and malaria prevention and treatment range between $1.3 billion and $1.6 billion a year over the 40-year period, that is, 2010–50.

Furthermore, there are some regional-scale studies assessing the cost of adaptation such as that in southern Africa (Van Rensburg and Blignaut 2002), according to which the total annual costs are estimated to be about $3800 million in 2025 (2000 price) or 3% of GDP per capita. These figures are more conservative but still comparable with the results of UNFCCC (2007) for malaria, where the valuation focuses on a somewhat larger timeframe and on all the African regions. The estimated costs appear to be low if compared with the total expected impacts of malaria in case of no adaptation (Epstein and Mills 2005), estimated in 1999 at about $90–270 billion without climate change.[5] This cost would be much higher in the event of climate change, which is expected to increase the relative risks of malaria, and clearly it would be well beyond the budgetary capacity of governments.

Some estimates from public health programmes

Kiszewski, Johns, Schapria, *et al.* (2007) estimate the financial resources required to implement a malaria programme to reduce the disease by at least 50% by 2010 and 75% by 2015 in the most affected malaria-endemic countries (in Africa, Asia and Middle East, and South America), as recommended by the World Health

[5] Estimates are based on the observed number of cases without interventions. Murray and Lopez (1996) have estimated around 36 million lost disability adjusted life years (DALYs) for malaria in 1999 in sub-Saharan Africa; two estimates of DALYs have been considered: the first equal to the per capita income and the second valued at three times the per capita income.

Assembly in 2005. The total financial resources needed from 2006 to 2015 to implement the recommended measures amount to $38–46 billion (on average, $3.8–4.6 billion per year). Among the different cost categories, the vector control costs are estimated to be the highest. The average annual costs (2006–15) are expected to be about $1.7–2.2 billion for Africa, about $1.9–4.6 billion for South East Asia, Western Pacific, and Eastern Mediterranean, and finally $212–235 million for Central and South America. The full range of costs is included in the assessment. These figures are comparable with the costs provided by the UNFCCC (2007), but the latter provide a much wider range to estimates. The authors have calculated a financial gap of 90%–95% between available domestic funding and resources needed. Actually, only 4.6% of the estimated annual resources could be covered by national funding in the African countries and only 9.2% in the countries of Asia, Oceania, and Americas. National governments are, therefore, unable to provide sufficient funding to meet the malaria targets established by MDGs.

According to Epstein and Mills (2005), the additional expenditure for malaria protection required in sub-Saharan Africa to achieve 40% population coverage for preventive measures and 50% coverage for treatment measures (according to the MDGs 2007) is in the order of $0.65–1.4 billion per year for prevention and $0.6–1.1 billion per year for treatment. These figures are higher than those estimated by Kiszewski, Johns, Schapria, *et al.* (2007), but similar to those provided by the UNFCCC (2007) under the unmitigated scenario for Africa regions ($2.6–7.2 billion). Another relevant study for malaria estimates the costs of selected malaria control programmes to meet the targets of the MDGs. For 95% population coverage, annual costs are estimated to be $468 million for western Africa and $442 million for southern and eastern Africa. These figures are much lower than those provided by Epstein and Mills (2005) and Kiszewski, Johns, Schapria, *et al.* (2007), as calculations exclude some cost items.

A more extensive study is that of Stenberg, Johns, Scherpbier, *et al.* (2007), which assesses the cost of implementing a health programme for children in 75 countries, according to the Fourth MDG requiring reduction in child mortality and morbidity by two-thirds by 2015. Total costs are estimated at about $52.4 billion during the period 2006–15, which corresponds to additional annual costs of $2.2 billion estimated in 2006 and $7.8 billion in 2015.

Available studies show that the magnitude of the costs of adaptation might be substantial and considerably higher than the current budgets allocated for health programmes by the governments in developing countries. Results confirm the need to increase funding resources for adaptation. There are, however, several problems in comparing findings across studies, mainly related to the types of interventions included in the analysis, the population coverage, and the targets achievable in terms of expected reduction in health impacts. Some degree of harmonization in reporting results would benefit the discussion considerably. Furthermore, there is a need to set up adequate databases of studies carried out and regularly updated using common methodological protocols. This would help policy-makers to prioritize and decide on the programmes to implement.

Finally, the review of existing literature reveals that the same methodology is used to estimate both preventive and reactive measures. The fact remains that reactive measures refer to the treatment of the disease to minimize the impact that has already occurred, whereas preventive measures should be applied to the vulnerable population at risk before the impact occurs to reduce or avoid the risk. In this study, we argue that prevention would play an important role in adaptation and could result in costs higher than those in reactive measures.

The methodological approach used in the current study is an attempt to address some of the limitations identified from the review of existing literature.

The methodological framework

This section describes the methodological framework used for estimating the health impacts due to climate change in India and the consequent costing of adaptation measures. The general approach applicable in the health context is presented here, whereas specific assumptions and adjustments are detailed in the sections dedicated to each disease. The analysis focuses on the following health outcomes, for which data was available: malaria, diarrhoea, and malnutrition.

The methodology consists of two main steps. These are: (1) the quantification of the health impacts attributable to climate change and (2) the assessment of preventive and reactive measures.

Quantification of the health impacts attributable to climate change

The health impacts attributable to climate change are estimated by comparing the number of cases in the absence of climate change to those under a climate change scenario (Campbell-Lendrum and Woodruff 2007).

The first scenario (with no climate change) is built on the projections of Mathers and Loncar (2006), who estimate the expected number of deaths in 2030 for the main world regions, based on the estimates of the WHO for 2002. These estimates are constructed for future scenarios of economic growth, based on the past correlation between increased wealth and mortality trends for specific causes. The first basic assumption is that economic development will positively impact the health status of the population. Second, it is assumed that developing countries will witness in the near future an economic growth comparable with that of richer countries, based on constant growth without crisis and negative impacts on economic development. The socio-economic variables considered by Mathers and Loncar (2006) include the GDP per capita, the "human capital" measured in terms of number of years of schooling by adults, and the impact of technology on population health measured with a time variable. The latter shows the positive effect of economic development on the health system. Population growth is estimated using the United Nations Population Division (UNPD) projections (taking into account fertility rate and migration) (UNPD 2003, 2005). Projections are reported for three scenarios: baseline, optimistic, and pessimistic, based on different assumptions of population growth and economic development.

As per these projections, the burden of disease in 2030 is expected to fall considerably in the South-East Asia Region (SEAR), when compared with 2008. In the baseline scenario, the number of deaths due to nutritional deficiencies is expected to decrease by 49%, whereas for diarrhoea and malaria, the decrease is projected to be about 31% and 33% on average, respectively. The figures, available only at regional level, have been used to estimate the projected burden of the three health outcome in India for 2030, using the average share of deaths reported from India in the SEAR.[6]

[6] Relevant data from WHO-SEARO. Details available at <www.searo.who.int>.

As projections of Mathers and Loncar (2006) are provided only for mortality, the initial step was to calculate the corresponding number of cases for each health outcome. For this purpose, time series mortality data for each health outcome, as reported by the Ministry of Health and Family Welfare of India for the period 1980–2005, is regressed with the number of cases and national GDP. Next, the derived coefficient of death is used to calculate the expected number of cases in 2030, under the assumption that the correlation between deaths and cases will show the same trend in the future. Finally, the projected number of cases for India is downscaled at state level using the percentage contribution of each state to the national burden of each disease.[7]

The second scenario (under climate change) is constructed taking into account the relative risks projected for 2030 (McMichael, Campbell-Lendrum, Kovats, *et al.* 2004) and applying them to the projected number of cases calculated from the estimates of Mathers and Loncar (2006), to obtain the additional number of cases attributable to climate change, under economic and technological development. The difference between the number of cases in the two scenarios (with and without climate change) represents the climate change attributable burden of disease.

Calculations are based on the Impact Fraction (McMichael, Campbell-Lendrum, Corvalan, *et al.* 2003), which is the climate change attributable fraction of the total disease burden, calculated by combining the population exposed and the relative risks (McMichael, Campbell-Lendrum, Corvalan, *et al.* 2003).

$$IF_i \frac{\sum P_i RR_i - 1}{\sum P_i RR_i},$$

where IF_i is defined as the impact fraction attributable to climate change for disease i; P_i is the proportion of the population exposed to the disease; and RR_i is the projected relative risk of the disease under different climate scenarios.

The relative risks are available at regional level and under different future climate scenarios between 2000 and 2030 which are as follows: (1) unmitigated scenario with no emission reductions,

[7] Ministry of Health and Family Welfare. 2009. Details available at <www.mohfw.nic.in>, last accessed on 1 March 2010.

(2) reduction in emissions with stabilization at 750 ppm CO_2 equivalent by 2210 (S750), and (3) reduction in emissions with stabilization at 550 ppm CO_2 equivalent by 2170 (S550).[8] The geographical regions include Africa, Americas, South East Asia, Europe, eastern Mediterranean, and western Pacific. The projected relative risks for SEAR region are reported in Annexure I. As India accounts for most of the population in SEAR region, it is a reasonable assumption to apply the corresponding relative risks for India.

The relative risks are estimated based on the assumption that climate change will develop slowly and constantly, with gradual changes in temperature and precipitation. The increased frequency of extreme weather, as well as critical thresholds, is not considered. Besides, specific actions that could be taken to adapt to climate change are not included in the risk factors.

Assessment of preventive and reactive measures

To reduce the health impacts of climate change, two main groups of measures can be adopted: preventive and reactive measures. The first aims at reducing the risk of occurrence of the disease (avoiding the impact), whereas the second reduces the impacts, once these have occurred. We hereby propose a two-step approach. In the first step, a set of preventive measures to be implemented in vulnerable areas to protect additional population at risk of climate change is identified and assessed. In the health context, a scenario where the additional risk of disease due to climate change is reduced back to the baseline level is, however, far from being realistic. Therefore, only a certain degree of reduction in the health impacts is assumed to be achieved with the proposed set of preventive measures. The assumed reduction is based on the effectiveness of the preventive measure, as documented in the public health literature (for example, Chow, Darley, and Laxminarayan 2007; Jamison, Breman, Measham, *et al.* 2006; Kiszewsky, Johns, Schapria, *et al.* 2007).

The second step consists of identifying and assessing a set of reactive measures (based on curative treatments) to cope with the residual health impacts and to reduce the corresponding damage.

---

[8] The following sections of the chapter refer to the mitigation scenarios as: unmitigated, S750, and S550.

The two phases of assessment, based on different methodological approaches, are discussed below.

Preventive measures

Preventive measures are targeted for vulnerable population, expected to be exposed to health-related climate change impacts. Some groups in the population are particularly vulnerable due to their physical, physiological or socio-economic conditions, such as children, pregnant women, low-income groups, people already in poor health, and so on. This study prioritizes, to some extent, the inclusion of most vulnerable groups, such as children and women in reproductive ages, while considering adaptation options.

The methodology for assessing preventive measures is based on five steps: (1) identification of vulnerable areas, (2) selection of preventive measures, (2) definition of population coverage for each measure, (4) estimation of adaptation costs, and (5) calculation of the expected reduction in the health impacts due to prevention, based on the effectiveness of measures.

In the first step, the geographical areas exposed to the increased risks are identified for 2030. These include areas currently exposed and expected to face a higher risk in the future, as well as new areas currently not exposed but projected to be at risk in the future. To capture the impact of economic development on population vulnerability in these areas, projections of Mathers and Loncar (2006) are used, based on the assumption that a future decrease in the burden of disease would be preceded by a proportionate decrease in the vulnerability of the exposed population.

In the second step, the main preventive measures are selected based on the national programmes currently being implemented in India while taking into account disease-specific guidelines given by the WHO and other initiatives such as Disease Control Priorities in Developing Countries (Jamison, Breman, Measham, *et al.* 2006). Although prevention includes non-structural measures applied to each person (such as impregnated bed nets), measures adopted to strengthen the health care system (such as additional human resources, staff training, office supplies, and mobility) and also structural measures such as water and sanitation programmes; only the first two categories are adopted in this study. These are based on the assumption that by 2030, India would have benefited from already established water and sanitation infrastructure, as economic development cannot be ignored while assessing adaptation.

In the third step, the population covered by prevention is identified, as the measures cannot be provided for the whole population living in a specific area. Different assumptions can be used for estimating this coverage, as reported in the public health literature. In this study, we use the targets stated by the WHO and those mentioned in the MDGs, also taking into account the population growth rate (Census of India 2001) in the identified areas.

In the fourth step, the unit cost of each measure, in terms of cost per person protected, is estimated and multiplied by the population covered, as follows.

$$\text{Cost}_{pi} = P_{pi} \times c_{pi},$$

where P_{pi} is the population covered by measure p for disease i, projected in 2030; c_{pi} is the unit cost of preventive measure p for disease i (cost per person protected); and Cost_{pi} is the total additional costs of prevention attributable to climate change.

The final step consists of estimating the expected reduction in the health impacts (effectiveness of the measure), which can be achieved with the proposed set of preventive measures. This information allows calculation of the residual number of cases, which will be in need of reactive measures. The level of population coverage and the type of measure are considered while evaluating the effectiveness of the interventions. The basic data for the evaluation is sourced from the public health literature (Jamison, Breman, Measham, *et al.* 2006).

A limitation of the study must be acknowledged here. The methodological approach for preventive measures has been applied only for malaria, given the availability of data and specific nature of the interventions involved. For diarrhoeal diseases, the main preventive measure consists of improved water and sanitation systems, which have been assessed separately and not as part of preventive measures. This is done assuming that by 2030, more adequate water and sanitation infrastructure would be in place in India as a consequence of economic development. Therefore, it is not the absence of basic infrastructure, which will matter in the near future, but it is the protection (climate proofing) and maintenance of these infrastructure, following the expected increase in frequency and intensity of flooding and other extreme weather events, which will matter. This would represent the true cost of preventive adaptation for diarrhoea, but it has been excluded from the analysis due to the complexity of the assessment and the lack of data. With regards to malnutrition, the main preventive measures are related to the improvement in agricultural production and to poverty alleviation, which are additional targets that need to be met and are mostly related to the country's development goals. Therefore, they have not been addressed in the health section.

Reactive measures

Reactive measures consist of curative treatments applied to the residual health impacts of climate change after prevention has occurred. First the measures to be implemented for each health outcome are selected, based on the national programmes currently implemented in India and taking into account the recommendations of the WHO.

The additional costs of reactive adaptation are then calculated by multiplying the number of residual cases attributable to climate change with the unit costs per case or per person treated.

$$Cost_{ri} = N_{Ri} \times c_{ri}$$

where N_{Ri} is the residual number of cases of disease i attributable to climate change in 2030; c_{ri} is the unit cost of curative treatment r for disease i (cost per case or per person treated); and $Cost_{ri}$ is the total additional costs of curative treatment. The residual impacts calculated are based on the additional number of cases projected in 2030 using the relative risks as described in the section "Quantification of the health impacts attributable to climate change" and also depend on the assumptions used for population coverage and effectiveness of preventive strategies.

The next section discusses, for each health outcome, the impact of climate change on population vulnerability in India, the main control programmes currently implemented, the measures selected for the cost assessment, and finally the results of the analysis.

Malaria

Among the various vector-borne diseases, malaria continues to pose a major threat to public health in India. The country is estimated to contribute more than 70% of malarial cases in the SEAR, and within the country, about 80% of the burden is confined to about 20% of population residing in high risk areas. The country's surveillance system has been reporting about 2 million annual malarial cases in the recent years, with *Plasmodium falciparum* cases accounting for almost half of the disease burden.[9] The Government of India has initiated several programmes and undertaken various measures to combat the menace of malaria in the country. Since the launch of the pioneering National Malaria Control Programme in 1953, several programmes have followed suit. Early case detection and prompt treatment, distribution of insecticide-treated nets (ITNs), selective indoor residual spraying (IRS), and capacity building for optimal utilization of the technical manpower are some of the core strategies followed by recent programmes. The Enhanced Malaria Control Project (1997–2005) that was implemented on a very large scale followed an integrated approach for malaria control. It was subsequently followed by the National Vector-borne Disease Control Support Project (2008/09–2012/13, under the aegis of the World Bank). The project supports the goals envisaged in India's National Health Policy, 2002 (Ministry of Health and Family Welfare 2002), and objectives of the National Rural Health Mission, 2005–12 (Ministry of Health and Family Welfare 2006). It aims to introduce the new National Drug Policy (Ministry of Health and Family Welfare 2007)[10] for malarial diagnosis and treatment in a phased manner in areas with high Pf burden.

Although a declining trend in the number of cases has been witnessed over the years, climate change is likely to impose an additional burden on malarial prevalence in India.

---

[9] WHO Malaria Country Profile India (1995–2007). Details available at <www.whoindia.org/LinkFiles/Malaria_Country_Profile-Malaria.pdf>.

[10] The National Drug Policy on Malaria is based on the WHO guidelines (WHO 2006) for prevention and treatment of malaria.

Climate and malaria

Malaria is probably the vector-borne disease most sensitive to long-term climate change (McMichael, Campbell-Lendrum, Corvalan, *et al.* 2003). The impact that changing climate conditions (temperature, precipitation, and humidity) may have on malarial transmission is evident from various studies that explore the association between climate variables and malarial incidence. For example, studies show that climate change may increase the risk of localized outbreaks of malaria in areas where the disease has been eradicated but vectors are still present, such as in Europe or the US (van Lieshout, Kovats, Livermore, *et al.* 2004). Regional climate changes have also been invoked as a major causal factor in the recent resurgence of malarial epidemic (*P. falciparum*) in the East African Highlands (Zhou, Minakawa, Githeko, *et al.* 2004). Analyses of updated temperature data for these regions establish a significant warming trend since late 1970s, with evidence that the magnitude of change is affecting transmission potential (Confalonieri, Menne, Akhtar, *et al.* 2007). In a time series study by Kovats, Bouma, Hajat, *et al.* (2003), the effect of El Niño southern oscillation (ENSO) on malaria epidemics in parts of South Asia and South America was found to be well established.

Similar analyses in the Indian context, however, remain limited. Table 2 provides a summary of findings from two key studies on climate change and malaria in India (Bhattacharya, Sharma, Dhiman, *et al.* 2006; Dhiman, Pahwa, and Dash 2008). These point towards likely spatial (geographical) and temporal (increase/decrease in duration of transmission seasons) shifts under future climate change scenarios. In a study by Bhattacharya, Sharma, Dhiman, *et al.* (2006), a set of transmission windows, in terms of different temperature and humidity ranges, is developed for the malarial parasite typical to Indian conditions, by analysing the present climate trends and corresponding malaria incidences. It is observed that though the broad transmission window for malaria in terms of temperature is between 15°C and 40°C, the number of days required for a parasite to complete its life cycle varies according to the number of days a particular range of temperature persists, provided the relative humidity remains conducive. Using these transmission window criteria, the most malarious regions in India are identified, both under baseline and greenhouse gas scenarios. The two studies report that some states will continue facing risk of malaria under climate change, whereas other states may see a reduction in the length of transmission season. On the other hand, some regions, which are not exposed today, are expected to come under the threat of malaria in the future due to climate change.

Table 2 Impact projections for malaria in India under climate change scenario

Reference	Current malarial impacts	Projected malarial impacts under climate change scenario
Bhattacharya, Sharma, Dhiman, *et al.* (2006)	Malaria most prevalent in the states of Madhya Pradesh, Jharkhand, Chhattisgarh, Orissa, West Bengal, and Assam	The paper establishes the likely influence of climate change on malaria using results of HadRM2 model, 1S92a scenario in 2050. Malaria likely to persist in Orissa, West Bengal, and southern parts of Assam, bordering north region of West Bengal in future as well. It may shift from the central Indian region to the south-western coastal states of Maharashtra, Karnataka, and Kerala. Also the northern states, including Himachal Pradesh and Arunachal Pradesh, Nagaland, Manipur, and Mizoram in the North East may become malaria prone.
Dhiman, Pahwa, and Dash (2008)	Malaria most prevalent in the states of Madhya Pradesh, Jharkhand, Chhattisgarh, Orissa, West Bengal, and Assam	Northern states, such as Jammu and Kashmir, Himachal Pradesh, Punjab, Haryana, Uttarakhand, and Uttar Pradesh, and the north-eastern states are likely to be more vulnerable to climate change. On the other hand, the southern states, that is, Andhra Pradesh, Karnataka, Tamil Nadu and Orissa, and some parts of Gujarat are likely to be less vulnerable to climate change.

Thus, the studies show that although malaria may shift to new regions under climate change scenario, there is a possibility of reduction in the incidence of the disease in other areas either due to a spatial shift or due to the shortening of the transmission season. These findings are nevertheless subject to a number of uncertainties related mainly to the presence of various environmental and socio-economic factors, other than climate, which influence the vulnerability of the population.

Impact projections and vulnerable population

The analysis makes use of the previously mentioned findings (based on Bhattacharya, Sharma, Dhiman, *et al.* 2006; Dhiman, Pahwa, and Dash 2008) for identifying populations at risk as well as for estimating future incidence of malaria in the identified vulnerable regions. Given that the incidence cases for malaria in India are available at the state level, the assessment makes use of states as spatial units for estimating costs. Furthermore, the study considers only the spatial dimension of impacts in terms of projected geographical shifts. Although some literature on the temporal impacts of malaria exists, wherein the duration of open transmission windows for malaria are predicted to change, this study does not factor in that component due to the lack of adequate data needed to carry out such a complex analysis as well as to minimize the uncertainty associated with modelled projections.

An inherent limitation of the impact studies is that they do not give information on differential risks that may exist within identified states. In the current study, only the population at high risk is considered for the cost assessment, based on the World Malaria Report (2008), which provides the percentage of population currently at high risk of transmission in endemic areas, for which prevention is planned. Furthermore, taking into account the WHO 2010 target (WHO 2008b), a scenario with at least 80% coverage of estimated vulnerable population is applied, assuming that India would have achieved this target by 2030.

Identification of measures and costs

The review of national control strategies along with WHO recommendations (WHO 2006) became the basis for the selection of preventive and reactive measures.[11] Annexure II lists out the major national malarial control initiatives that have been implemented over the years.

---

[11] Information mostly sought from National Vector Borne Disease Control Programme (NVBDCP), Government of India. National Vector Borne Disease Control Support Project (2008–13), National Project Implementation Plan, Ministry of Health and Family Welfare, New Delhi, June 2008. Details available at <nvbdcp.gov.in/Doc/PIP.pdf>.

For assessing the cost of prevention, IRS and distribution of long-lasting insecticide nets (LLINs) are considered, as these form primary vector control components of the malarial control programmes in India. Although currently, ITNs are in use, national programmes aim at promoting the use of LLINs in the future. Because information on stratification of states/districts with respect to malarial risk is unavailable, it was impossible to delimit areas or populations, which would be targeted for selective interventions (IRS or LLINs). Therefore, for the purpose of this study, both interventions are extended to the entire identified target population. Cost of intermittent preventive treatment during pregnancy is also accounted for, based on the assumption that the number of pregnancies per year in malaria endemic regions is about 1.09% of the population at risk (UN 2003). Finally, some measures, as detailed out by the National Vector-borne Disease Control Support Project, pertaining to the strengthening of existing health infrastructure are also assessed. These include cost of increasing human resources, transport facilities, office supplies and other support, and most importantly, training of staff.

For reactive measures, both diagnosis and treatment are included in the analysis. Currently, laboratory microscopy is used for diagnosis, but health agencies and volunteers running fever treatment programme in inaccessible areas are provided with rapid diagnostic kits (RDKs). At present, different treatment policies are applied in different parts of the country based on the chloroquine resistance level. While artemisinin-based combination therapy (ACT) is used as first line treatment for all confirmed *P. falciparum* cases in chloroquine-resistant areas, positive cases of *Plasmodium vivax* are treated with chloroquine. It is assumed that 1% of the fevers among the residual cases would progress to severe malarial cases (WHO 2007), thus following a separate treatment protocol. RDK and ACT storage and distribution costs are also included in the analysis.

Unit costs for all products are measured in 2004 dollar price and sourced from the WHO (WHO-RBM-UNICEF-PSI-MSH 2004). The unit costs for strengthening the health care system are also derived from WHO-RBM-UNICEF-PSI-MSH (2004) and WHO (2007) in conjunction with implementation documents of national malarial control programme.[12] The study ignores the costs of programmes that are primarily implemented in non-health sectors, such as those aimed at better water and sanitation but which might have an influence on malarial incidence.

Results

The projections by Mathers and Loncar (2006) reflect varying degrees of reduction in malarial incidence by 2030 when different levels of development are taken into account. However, climate change can be seen as an additional stressor for human health, which imposes an additional burden on the financial resources supporting the health infrastructure of the country. This would be especially true for states that have been projected to have a climate conducive to malaria under a greenhouse gas scenario (Bhattacharya, Sharma, Dhiman, *et al.* 2006; Dhiman, Pahwa, and Dash 2008). Tables 3 and 4 show major findings of the study. The additional annual cost of adaptation for malaria in India in 2030 ranges from $71 million (in the optimistic scenario) to $179 million (in the pessimistic scenario) under unmitigated emissions. The difference amongst the three scenarios can be ascribed to the differential assumptions related to economic and technological development, GDP, and population growth. As for the mitigation scenarios, the cost difference is not too high[13] when compared with the unmitigated emission scenario with possible cost reduction ranging from 2% in the pessimistic scenario to 5% in the optimistic scenario. In each case, the share of preventive costs was found to be higher than that of the reactive costs, as prevention entails a larger coverage of population (Table 5).

The results show that most of the malarial burden under the climate change scenario would be concentrated in the north-eastern part of India,[14] along with some coastal states, both on the West Coast (Maharashtra, Karnataka, and Kerala) and the East Coast (Orissa and West Bengal). Some of these states, such as Orissa, Karnataka, West Bengal, and the north-eastern states, are

---

[12] Same as footnote 11.

[13] This can be explained by the small difference in the relative risk factors between the emission scenarios (McMichael, Campbell-Lendrum, Kovats, *et al.* 2004).

[14] The north-eastern region of India includes seven states of Assam, Mizoram, Sikkim, Meghalaya, Nagaland, Manipur, and Arunachal Pradesh.

Table 3 Additional cost of adaptation (preventive and reactive) for malaria in India, by state, under unmitigated emission scenario

States at risk of malaria	Development scenario		
	Optimistic	Baseline	Pessimistic
Arunachal Pradesh	237	392	595
Assam	5 695	9 398	14 242
Jammu and Kashmir	2 151	3 548	5 376
Himachal Pradesh	1 197	1 976	2 993
Kerala	5 841	9 638	14 603
Meghalaya	490	810	1 228
Mizoram	186	308	466
Nagaland	414	683	1 035
Maharashtra	21 303	35 152	53 263
Karnataka	10 616	17 517	26 542
Manipur	500	742	1 125
Orissa	7 243	11 961	18 132
West Bengal	15 881	26 205	39 707
Total cost	71 754	118 330	179 307

Note For 2030 in '000 $ at 2004 prices, without discounting

Table 4 Additional cost of adaptation (preventive and reactive) for malaria in India, by state, under mitigation scenarios (S550 and S750)

States at risk of malaria	Development scenario		
	Optimistic	Baseline	Pessimistic
Arunachal Pradesh	221	373	571
Assam	5 398	9 092	13 924
Jammu and Kashmir	2 043	3 441	5 269
Himachal Pradesh	1 137	1 916	2 933
Kerala	5 549	9 345	14 310
Meghalaya	462	778	1 191
Mizoram	176	297	454
Nagaland	393	662	1 013
Maharashtra	20 230	34 071	52 172
Karnataka	10 079	16 975	25 994

Contd...

Table 4 *Contd...*

States at risk of malaria	Development scenario		
	Optimistic	Baseline	Pessimistic
Manipur	427	720	1 102
Orissa	6 836	11 516	17 637
West Bengal	15 076	25 392	38 882
Total cost	68 027	114 578	175 452

Note For 2030 in '000 $ at 2004 prices, without discounting

already highly malarious with an Annual Parasite Index higher than five. The state of Orissa alone contributes to one-fourth of all malaria cases currently reported in India (Kumar, Valecha, Jain, *et al.* 2007; Lal, Sonal, and Phukan 2000). The projected increase in temperature and relative humidity in the future is likely to expand the lower and upper thresholds of malarial transmission window, bringing in new areas under enhanced malarial risks (such as high altitude areas of Jammu and Kashmir) or same areas under longer transmission spells (such as in case of Orissa and Assam).

Table 5 Summary of total additional cost of adaptation (preventive and reactive) for malaria in India under different development and emission scenarios

Development scenario	Emission scenario					
	Unmitigated			S550 and S750[a]		
	Preventive	Reactive	Total	Preventive	Reactive	Total
Optimistic	71 496	208	71 704	67 921	105	68 026
Baseline	117 968	361	118 330	114 394	183	114 576
Pessimistic	178 740	569	179 308	175 165	287	175 452

[a] No difference between relative risks is reported for the two scenarios, S550 and S750.
Note For 2030 in '000 $ at 2004 prices, without discounting

Table 6 reports an estimate of the current expenditure on malaria control in India. It is, however, essential to note that the total fund indicated is much lower than what is currently required, because the existing coverage of preventive measures is inadequate[15] and most of the cost burden is borne privately by the patients. Nonetheless, as the results show, climate change will add to the financial burden, requiring the government to expend more and probably demanding additional aid from external funding sources. Although current expenditures show that the share of preventive measures is less than that of the reactive measures,[16] it is interesting to see from the results of the analysis that a more effective coverage of the vulnerable population by preventive measures may actually reduce the morbidity and mortality burden caused by malaria and the associated costs to the society.

Table 6 Government and external funding for malaria control in India (2007)

Source of funding (includes IRS, ITN, diagnosis, and treatment)	Expenditure ($)
Global fund for AIDS, tuberculosis, and malaria	15 727 050
World Bank	29 500 000
Government expenditure	99 010 000
Total	144 237 050

IRS – indoor residual spraying; ITN – insecticide-treated nets

Source WHO (2008b)

Diarrhoeal diseases

According to the 2004 Global Burden of Disease update (WHO 2008a, b), diarrhoeal diseases are the fifth leading cause of death in the world, contributing an estimated 3.7% to global mortality, whereas in the low income countries, it is the third most common cause of death, accounting for about 6.9% of the total deaths. Globally, diarrhoeal diseases still rank second among the causes of death in children below the age of five (UNICEF–WHO 2009),

---

[15] In 2007, operational indoor residual spraying (IRS) coverage was 40%; operational insecticide-treated nets (ITN) coverage was less than 5%; and no intermittent preventive treatment (IPT) of malaria was provided to pregnant women.

[16] In 2007, of the total expenditure on interventions for malaria in India, approximately 54% was spent on treatment (including diagnosis) and approximately 45% on prevention (IRS and ITN) (WHO 2008b).

contributing about 16% to the total mortality in the world and about 20% in India. Because diarrhoeal incidence is influenced by a number of factors, both environmental (including climate) and socio-economic, actions to reduce diarrhoeal burden must encompass a wide range of preventive (structural and non-structural) and reactive measures. A number of interventions that have direct or indirect impact on diarrhoeal incidence are being implemented in India. The programmes are spread across a number of departments under the purview of various ministries such as that for health, water, sanitation, and urban and rural development. In spite of these efforts, diarrhoea still remains a big challenge for India, both in terms of mortality and morbidity, especially among children (Third Round of National Family Health Survey (f-3), 2005/06 data in International Institute for Population Sciences [IIPS] and Macro International 2007). Climate change causes alterations in temperature, humidity, surface water, and groundwater availability, and deteriorating quality of water can be expected to aggravate water-related diseases in India, where a large part of the population does not have access to regular supply of safe drinking water, sanitation, adequate nutrition, and effective health care.

Climate and diarrhoeal diseases

Available research also suggests a significant future increase in heavy rainfall events in some regions of India, whereas others face increased drought conditions (GoI 2004; IPCC 2007). There is already enough evidence of association between diarrhoeal diseases and climatic variability on both temporal and spatial scales in different regions of the globe (Checkley, Epstein, Gilman *et al.* 2000; Pascual, Rodo, Ellner, *et al.* 2000; Purohit, Kelkar, and Simha 1998; Rodó, Pascual, Fuchs, *et al.* 2002; Singh, Hales, de Wet, *et al.* 2001). Projected climate change resulting in warming, increased frequency of extreme events, sea level rise, and melting of glaciers will undoubtedly affect the water balance and water quality in different parts of the country, which will consequently affect human health. Preliminary assessments carried out for India's first National Communication (NATCOM) to the UNFCCC

(GoI 2004) show that under the IS92a[17] scenario, the severity of droughts and intensity of floods in various parts of India are likely to increase. A general reduction in the quantity of available run-off under the IS92a scenario, for the period 2041–60, is to be expected, which would affect the water availability in many a river basins of the country. River basins like Sabarmati and Luni may experience acute water scarcity, along with basins of Mahi, Pennar, Sabarmati, and Tapi, which are likely to experience constant water scarcity and shortage (Gosain, Rao, and Basuray 2006). In other basins, such as in Mahanadi, increased rainfall intensity may lead to higher run-off, thereby reducing recharge. Thus, along with surface water, climate change is likely to affect groundwater resources due to changes in precipitation and evapotranspiration. Intrusion of saline water into coastal aquifers due to sea level rise may lead to coastal populations facing water scarcity.[18] According to the first NATCOM, a 1 m sea level rise is projected to displace approximately 7.1 million people in India. An increase in the frequency of severe cyclonic storms is also likely under climate change scenario, which enhances the vulnerability of coastal populations to direct (injury and loss of lives) and indirect health impacts (related to water, sanitation, displacement, and so on).

Impact projections and vulnerable population

Unlike the malarial analysis, population that would face increased risk of diarrhoea in the future could not be ascertained, primarily, due to lack of impact studies in this regard as well as the complexity of interaction between contributing factors that make such an impact analysis difficult. The projection of additional diarrhoeal burden is, therefore, based only on the additional number of diarrhoeal cases estimated under climate change scenario using data on relative risks. This makes the estimated costs an underestimation of the actual additional cost that climate change imposes, as it is exclusive of preventive costs.

---

[17] IS92a scenario has been widely adopted as a standard scenario for use in impact assessments. It assumes moderate population growth (reaching 11.3 billion by 2100) and high economic growth (2.3% per annum between 1990 and 2100), high fossil fuel availability, and eventual phase out of nuclear power.

[18] For more information on coastal vulnerability hotspots, refer to chapter, "Coastal zones".

The analysis is targeted at children of up to five years, as this age group accounts for the major share of diarrhoeal burden in India; the group is also the main target for the current governmental programmes for control of diarrhoea (as per NFHS-3 in IIPS and Macro International 2007).

Identification of measures and costs

The review of national control programmes, along with WHO recommendations for management of diarrhoea (WHO-UNICEF 2004; WHO 2005), became the basis for selection of preventive and reactive measures. Annexure III lists some of the initiatives that constitute the core of diarrhoeal control and child health care programmes in the country.

Breastfeeding promotion, cholera immunization, oral rehydration therapy, vitamin A, and zinc supplementation are the main measures that were considered for the cost assessment on the basis of review of current clinical management guidelines followed in the country, along with WHO recommendations. The costs of programmes aimed at improving the water and sanitation situation, which can be considered as the main structural preventive measure in case of diarrhoea, were excluded from the cost estimation, as the scenarios of economic development adapted from Mathers and Loncar (2006) are assumed to be inclusive of these costs. The costs estimated are, therefore, the costs of managing only the additional climate change attributable cases of diarrhoea, assuming that by 2030, infrastructure required for water and sanitation improvements will be already in place. Unit costs for the various interventions (in 2001 dollar price) are taken from Disease Control Priorities in Developing Countries Project (Keusch, Fontaine, Bhargava, *et al.* 2006), which provides range of prices for diarrhoeal control measures.

Results

The results of the study (Table 7) show that annual adaptation costs needed for the management of climate change attributable cases of diarrhoea in 2030 range from $57 million to $79 million in the optimistic scenario to $218 million to $303 million in the pessimistic scenario, depending on the level of mitigation achieved. Comparison between the costs in unmitigated emission and the

Table 7 Additional cost of adaptation for diarrhoea in India under various scenarios of development and emissions

Development scenario	Emission scenario		
	Unmitigated	S750	S550
Optimistic	79 616	65 050	57 367
Baseline	177 037	144 647	127 564
Pessimistic	303 575	248 034	218 742

Note For 2030, in '000 $ at 2004 prices, without discounting

two mitigation scenarios shows that a cost reduction of an average 18% can be achieved if emissions are stabilized at 750 ppm and an average 28% in case of a 550 ppm mitigation scenario. However, it must be emphasized here that the health adaptation cost estimated for the various scenarios assumes that the infrastructure for water and sanitation would be already in place, presuming that the country not only meets the MDG targets of halving the number of people without access to water and sanitation, but also surpasses it by 2030. If in case adequate adaptation in the water and sanitation sectors does not occur, the burden and cost of managing additional cases of diarrhoea would be much higher than what is indicated (Box 1).

Under a future climate change scenario, expansion and improvements in water, sanitation, and most importantly, health care facilities must be prioritized in states that are currently having high rates of diarrhoeal incidence or which are projected to experience water deficits.[19] The impact of extreme events such as floods and cyclones also need to be considered while planning basic infrastructure. Early warning systems, disaster management strategies, and emergency responses play a vital role in preventing outbreaks or epidemics following a climate disaster. Currently, coastal states seem to be particularly affected with diarrhoea, with four coastal states of West Bengal, Andhra Pradesh, Maharashtra, and Karnataka contributing to almost half of the annual diarrhoeal incidence in the country. Planning of coastal protection measures, along with adaptation to hazards, such as rising sea levels, fresh

---

[19] Rajasthan and Gujarat that are dependent on river basins of Tapi and Sabarmati are projected to experience water scarcities under climate change scenario (refer to the section "Impact projections and vulnerable population)."

Box 1 Improvements in water and sanitation infrastructure in India

As provision of safe drinking water, adequate sanitation, and a clean environment are vital to reduce incidence of water-borne diseases and deaths, the high burden of diarrhoea draws attention towards the status of water and sanitation in the country. As per the country's Eleventh Five-year Plan for the period 2007–12 (Planning Commission 2008), currently, drinking water makes up only less than 1% of the total water demand. With demand for water growing in every sector (domestic, agriculture, industry, and others), there is ever mounting stress on India's finite and fragile water resources. Rural areas and peri-urban areas are first to get affected by water scarcities as systems of regulated water supply are either poorly managed or absent, and this, combined with low income levels, lack of basic awareness, and poor nutritional intakes, leads to high rates of infectious diseases, especially among children. Statistics reveal large inequalities in water supply and sanitation coverage in different regions of the country. Figures show that only about 74% of rural habitations have access to protected drinking water sources and very few states such as Tamil Nadu, Delhi, and Uttar Pradesh have achieved full coverage of rural habitations, whereas others are still lagging behind (MoF 2008). Similar disparity can also be seen in the per capita water availability. On the urban front, 91% of the urban population has access to water supply facilities, but distribution networks are poorly maintained, leading to water wastage and contamination. The access to sanitation facility is even poorer. As per the latest Census data (Census of India 2001), only 36% of the total population has toilets within or attached to their houses. However, in rural areas, only 21.9% of the population has toilets within or attached to their houses, putting the total sanitation coverage in the country at about 49% (Planning Commission 2008). This impact of disproportionate access to water and sanitation facilities gets reflected in the differences in annual diarrhoeal incidence at the state level.

Given the relevance of this issue, it is interesting to provide to policy-makers some figures on what could be the cost of improving the current water and sanitation infrastructure in the next years in India. Estimates are based on the study of Hutton and Haller (2004), which provides the annual cost per capita (based on the entire population) for different interventions required for improving water and sanitation in South East Asia region by 2015. Taking into account the Millennium targets recommendation to half the proportion of

Contd...

Box 1 *Contd...*

people with no access to improved water and sanitation facilities by 2015, and considering the predicted population growth in India under different IPCC scenarios, the total annual costs of achieving this target for India are estimated around $2.6 billion in scenario A2r, $2.5 billion in scenario B1, and $2.4 billion in scenario B2. To bring the entire population under water supply and sanitation services, the total annual costs would rise to $5.2 billion in scenario A2r, $5.1 billion in scenario B1, and $5 billion in scenario B2.

water depletion, and cyclones, becomes important for safeguarding the health of coastal communities.[20]

Annexure IV provides downscaled adaptation cost figures at the state-level based on the current distribution of total diarrhoeal burden amongst the states of India. It must be mentioned that different states in the country enjoy different levels of health-care services, which is bound to play an important role in the existing differential disease incidence rates.

Malnutrition

As per the Global Hunger Index study (von Grebmer, Fritschel, Nestorova, 2008), India is amongst the countries that have the highest prevalence of underweight children under five years of age (almost 40%), and only modest progress has been seen over last decades through efforts aimed at reducing the malnutrition burden, including micronutrient deficiencies (Gragnolati, Shekar, Gupta *et al.* 2005). The state of micronutrient deficiencies is also dismal. Although the prevalence of anaemia among women in India varies widely among different states, it is still widespread, with about 56% women suffering from this condition and only 22% pregnant women consuming iron or folic acid supplements. Iodine deficiency, which can lead to a number of disorders, is also common in India. According to NFHS-3 2005/06 (IIPS and Macro International 2007), only 51% households in the country use adequately iodized salt.

[20] Refer to chapter, Coastal zones, for more detailed assessment of impacts and adaptations to climate change in coastal regions of India.

It has been observed that disorders related to inadequate and improper nutrition are substantially higher in rural areas of India than in urban areas.

As inter-annual, monthly, and daily distribution of climate variables (primarily temperature, precipitation, and humidity) play a fundamental role in most processes that drive productivity in agricultural, forestry, and fisheries systems, any change in climatic determinants can not only lead to adverse impacts on food security and nutrition, but also essentially affect well-being of millions deriving livelihood and income from these.

Climate and malnutrition

Climate change is expected to worsen the problem of malnutrition in India and hamper progress achieved under nutritional programmes, as agricultural processes and the availability of food around the world get affected. Although in some areas, changes in rainfall and temperature may actually increase the potential to grow crops and, in turn, increase food security, in most areas, climate change is expected to cause more droughts and floods, which would be damaging to the crops. The projected beneficial nature of the so-called CO_2 fertilization effect must also be considered in light of the accompanying changes in air temperature, moisture availability, survival and distribution of pest populations, frequency and intensity of inter- and intra-seasonal droughts and floods, soil organic matter transformations, soil erosion, decline in arable areas due to submergence of coastal lands, and socio-economic implications on farming systems (Mall, Singh, Gupta, *et al.* 2006; Parry 2007). An important dimension of the problem would be to see how global trade in agricultural products would change under different climate scenarios. As agricultural production increases in the developed countries at temperate latitudes and decreases in the developing countries of the tropics, current balances of trade will be altered because agricultural exports make up a large portion of the GDP of many developing countries. According to Parry, Rosenzweig, Iglesiasc, *et al.* (2004), even if the global production remains stable, regional differences in crop production would intensify, leading to increases in prices and risk of hunger in the less developed countries, which may not be able to afford imports of food commodities.

While the lack of sufficient income to purchase food is a major factor causing households to face food insecurity, hunger itself contributes to poverty by lowering labour productivity, reducing resistance to disease, and undermining educational achievements (FAO and World Bank 2001). This vicious cycle initiated between "food insecurity–malnutrition" and "livelihood loss–poverty" may impede economic progress in a developing country like India. In some areas where livelihood choices are limited, decreasing crop yields may lead to famines, or where loss of landmass in coastal areas is anticipated, migration might be the only solution (ADB and International Food Policy Research Institute 2009).

Climate change can also be expected to reinforce the association between malnutrition and some infectious diseases (Caulfield, Richard, and Black 2004). While malnutrition can increase susceptibility to infections by inducing alterations in hosts' immune function, infectious diseases adversely affect nutritional status by reducing an individual's capacity for food intake and nutrient absorption (Brown 2003). Some studies (Edirisinghe 1986) show that deficiencies in vitamin A, zinc, iron, and other micronutrients are responsible for a substantial proportion of malaria morbidity and mortality, as these nutrients are vital for building natural resistance against malarial infection. A reciprocal relationship has also been postulated between diarrhoea and malnutrition in children, with diarrhoea leading to nutrient loss and malnutrition predisposing diarrhoea. A study in the rural area of Bangladesh (Black, Brown, and Becker 1984) found that the duration of diarrhoea increased progressively as nutrient status indicators worsened. According to Guerrant, Schorling, McAuliffe, *et al.* (1992), malnutrition can result in a 37% increase in frequency and 73% increase in duration of diarrhoeal symptoms, accounting for the doubling of the diarrhoea burden in terms of symptoms days in malnourished children. It may thus be recommended that nutrition programmes be integrated into existing treatments for malarial and diarrhoeal diseases, because malaria and diarrhoea are already much widespread in India. With respect to diarrhoea, some micro-nutrients have been included in the present study, whereas nutritional programmes have been assessed only for malnutrition, to avoid double counting issues and due to the existing complexity to disentangle the impacts related to the different health outcomes.

Impacts and vulnerable population

For this study, two broad categories were identified within malnutrition: protein-energy deficiency and micronutrient deficiencies with respect to iodine, vitamin A, and iron. Findings from National Family Health Survey-3 (IIPS and Macro International 2007) are reviewed for selecting the most vulnerable groups. Although protein-energy malnutrition has been addressed in children between ages 0 and 14, cases related to iodine and vitamin A for all age groups have been accounted for. For anaemia caused by iron deficiency, the most vulnerable groups, as suggested by the national survey, are children (0–4 years), pregnant women, and non-pregnant women of reproductive age.

As in the case of diarrhoea, the population expected to be at risk of malnutrition in the different states of India could not be assessed because of the dearth of impact studies in this context. Furthermore, preventive actions in this context are mainly related to enhancing agricultural productivity and reducing poverty. These impacts have not been assessed as part of the health adaptation as they involve development targets in the context of MDGs, which would imply cross effects between different sectors. Therefore, the assessment is only based on the projected additional cases of malnutrition and micro-nutrient deficiencies attributable to climate change (based on the relative risks), expected in India by 2030.

Costs of interventions

The main programme of the Government of India to provide nutrition for children is the Integrated Child Development Services (ICDS) programme. Supplementary feeding, growth monitoring, immunization, micronutrient supplementation, early health care, and preschool education are some of the key components of the programme. In particular, promoting awareness amongst mothers on breastfeeding immediately after birth, continued feeding during illnesses like diarrhoea, and supplementation with selected micronutrients form a part of an important strategy aimed at controlling nutritional complications in children. Pregnant and lactating women are also entitled to receive a variety of services through the ICDS programme. The Government of India recommends that preschool age children must be provided with vitamin A supplements to prevent vitamin A deficiency disorder.

For the purpose of cost estimation, the following measures are considered: (1) for protein-energy malnutrition in children between ages 0 and 14, unit costs of nutritional programme and "child survival programme" with nutrient component and (2) for micronutrient deficiencies unit costs of supplements for iodine, vitamin A, and iron deficiency. Unit costs were taken from Disease Control Priorities in Developing Countries Project (Caulfield, Richard, Rivera, *et al.* 2006), which provides range of prices for strategies for management of malnutrition and micronutrient deficiencies.

Results

The analysis reveals that additional annual adaptation cost for addressing malnutrition may range from about $16 million in the optimistic mitigation scenario to about $63 million in the pessimistic scenario without mitigation (Table 8). On average, about 20% reduction in additional financial resources needed for adaptation can be achieved if proper mitigation strategies are planned and invested in. However, the costs estimated here might have a degree of underestimation associated, as only nutritional deficiencies in select age groups and sex are considered. Also, it is exclusive of measures that would be needed to reduce the vulnerability of the population such as soft measures for education and awareness creation. However, there might also be some degree of double counting as cost incurred in some of the measures, such as breastfeeding promotion and vitamin A supplementation, are also accounted for in the analysis on diarrhoea. Because evidence suggests that the occurrence of diarrhoea and malaria may lead to nutritional losses, especially in case of children, it can be reasonably assumed that the strategies implemented to achieve malarial or diarrhoeal prevention and control may also lead to a reduction in cases of malnutrition and associated costs. The reverse may also be true. These costs must also be seen in the light of the adaptation costs in other key sectors, such as agriculture and water, which are the major determinants of food security and livelihood opportunities.

Table 8 Additional cost of adaptation for malnutrition in India under various development and emission scenarios

Development scenario	Emission scenario	
	Unmitigated	S550[a]
Optimistic	19 859	16 009
Baseline	36 793	29 660
Pessimistic	63 658	51 316

[a] Scenario S750 is not reported due to the small difference in the cost assessment with the unmitigated scenario.
Note For year 2030, in '000 $ at 2004 prices, without discounting

The lack of sufficient data on malnutrition incidence, by state, limits the downscaling of total adaptation costs to state level. However, there is a reason to believe that both future disease burden and cost of managing the condition will fall inequitably on different states, considering the current gap in the performance of different states with respect to child underweight and child mortality.[21] The India State Hunger Index (Menon, Deolalikar, and Bhaskar 2009) constructed for 17 major states of India, capturing three dimensions of hunger (inadequate calorie intake, child malnutrition, and child mortality), indicates substantial variability among states in India, with the index scores ranging from 13.6 to 30.9. Of the 17 states, 12 analysed fall in the "alarming category" and 1 in the "extremely alarming" category. A strong association between the index score and poverty level in many states has also been highlighted in the study along with a lack of correlation between trends in economic growth (measured in terms of the net state domestic product) and trends in the rates of hunger. This indicates that poor economic growth may not always be the driver of hunger and undernutrition in a state. Thus, equitable economic growth, accompanied with direct investments targeting poverty alleviation and nutritional programmes, is necessary to raise the levels of nutritional performance in a state or a region.

[21] The results of NFHS-3 (2005/06 data) give state-wise rates for underweight and calorie undernourishment in children (IIPS 2007).

Knowledge gaps and limitations

We would argue that only by a detailed bottom-up analysis of this kind can one obtain credible estimates of the costs of adaptation to climate change in the health area. To date, such studies are rare, and this is the first national-level estimate we are aware of. Nevertheless, although the study attempts to address some of the major lacunae identified in previous assessments on health adaptation costs, there still exist several issues that require further analysis.

One of the major limitations of the study stems from the poor information base on the potential future health impacts of climate change on India, especially at the regional level. Accurate cost estimation requires robust impact assessments, so that populations most vulnerable are delimited, additionality attributable to climate change is estimated, and costs of suitable preventive measures are calculated. Given the scant information available on impact projections and lack of consistent long-term datasets, the study limits its analysis within the scope of available information. It thus estimates the impact of climate change on three health outcomes only, that is, malaria, diarrhoea, and malnutrition. Only for malaria, the cost assessment distinguishes between preventive and reactive measures, as the population targeted with the former has been identified using results from impact projections, whereas the latter is based on residual malarial morbidity, assuming that effectiveness of preventive measures cannot be complete. Other important climate-sensitive impacts such as those from thermal stresses and rainfall extremes have been excluded from analysis. However, because the three outcomes included in the analysis constitute the major health challenge in India, from a climate change perspective, the cost estimated reflects much of the additional burden imposed

As health concerns are intricately linked to impacts and developments in other key sectors (such as water resources, agriculture, infrastructure development), overlaps between sectoral adaptation measures require careful consideration while undertaking a cost assessment. However, delineating the boundaries between different sectors and defining implications of cross-sectoral adaptations are part of a complex process. This study recognizes the overlapping nature of the sectors and makes every attempt to minimize duplication of costs.

Often, in the adaptation literature, debate surrounding options between soft adaptation and hard adaptation come into picture. Most of the earlier studies on cost of adaptation focus on estimation of the cost of hard structural adaptation measures with less emphasis on the soft modes of adaptation, even though it is recognized that soft adaptation options are more cost-effective and avoid the need for huge physical investments. The complexity involved in capturing the soft adaptation measures is in understanding the nuances of existing local institutions and community-level actions and evaluating these. The present study also confronts similar set of difficulties in estimating the cost of soft adaptation measures. Another concern related to the adaptation costs is the emphasis on public adaptation with less attention to the private modes of adaptation. This reason for such preference in most adaptation cost estimates is due to the difficulty in identifying the cost and degree of autonomous adaptation at the individual, household, and firm levels. The current study also includes only planned adaptation measures and resultant costs. However, planned and autonomous adaptation measures can be considered to be complementary to each other. Carrying out rigorous cost-effectiveness analysis to identify the most preferred adaptation options is another crucial aspect. Lack of data largely constraints this kind of analysis, making it difficult to comment on the optimal combination of hard and soft options; autonomous and planned actions; and preventive and reactive measures. In this regard, the current study is similar to preceding studies in looking only at the cost numbers without going into the detailed analysis of the cost-effectiveness of adaptation options. However, the measures included in the analysis have a basis in the health and adaptation literature and draw heavily from recommendations in existing health policies and programmes.

Another aspect of cost estimation process, which the current study does not consider, is related to the distributional impacts of post adaptation phase. Often the market can solve the problems of climate through adaptation, but the distribution of market benefits is often a challenge where people having low bargaining power stand to lose. True distributional impact of adaptation requires an understanding about the "public" and "private" good nature of adaptation options. Related concern in the estimation of health adaptation costs is the "adaptation deficit". Costs of adaptation can be studied in two ways, with and without adaptation deficits. The complexity involved in distinguishing between development deficit and adaptation deficit poses a major roadblock for true cost estimation. Differentiating development deficit from the adaptation deficit is more difficult in developing countries like India. However, the current study attempts to internalize the impacts of development targets, such as those envisaged in the MDGs, and then, accordingly, estimate costs for non-structural measures, but with the caveat that there still exists the difficulty in separating the adaptation and development deficits. Estimation of the extent of the residual demand is also an important component of the adaptation cost estimation exercise, which needs to be addressed.

Summary and conclusion

As already discussed in the preceding sections, climate change imposes a considerable additional burden on society, which falls disproportionately on the vulnerable poorer groups of population having lower adaptive capacity. The present study, with its limitations, furthers the understanding on the health impacts of climate change in the Indian context and provides cost estimates imposed by the additional burden of malaria, diarrhoea, and malnutrition under a scenario of climate change. Results of the analysis show that the additional annual adaptation costs under different scenarios of development are in the range of $171–546 million in the unmitigated scenario and $141–445 million in a scenario of stabilization at 550 parts per million by volume (ppmv), for 2030 (Table 9). A stabilization scenario allows for a possibility of achieving approximately 15%–18% reduction in estimated annual additional costs of adaptation in the health sector, depending on the development scenario.

Total annual costs of adaptation for malaria and diarrhoea in India are estimated to range from $152 million to $483 million in the unmitigated scenario. These figures are comparable with, but lower than, those estimated by the World Bank (2009) study, where the cost of adaptation for the two diseases in the whole South Asia is estimated to be about $850 million a year. The cost of treating malnutrition is lower compared with the cost of adaptation estimated for malaria and diarrhoea. This is due to the fact that this study estimates specifically the cost of measures related to the health sector while excluding more general multiple-benefits interventions, such as improvements in crop yields and reduction of poverty, which are expected to face important additional costs to be supported by other sectors.

Table 9 Additional cost of health adaptation to climate change in India under various development and emission scenarios

Development scenario	Emission scenario	
	Unmitigated	S550
Optimistic	171 333	141 403
Baseline	332 160	271 800
Pessimistic	546 334	445 510

Note For 2030, in '000 $ at 2004 prices, without discounting

If we look at the current health status of India, financial resources allocated for the sector clearly stand out to be inadequate in effectively benefiting the masses. The rapid economic growth achieved by the country in the last two decades has supported some improvements in health status in terms of reduced infant mortality and increased life expectancy, but the rate of improvements make achievement of many of the MDGs targets appear a far-fetched goal. Statistics for 2006 show that India's total expenditure on health was only about 3.6% of GDP, which was lower than the average for "low income" group in that year[22] and much lower than many other countries that have comparable per capita GDP. Also, private health spending is far greater than public health spending. Only about 25% of the total expenditure is accounted for in general government expenditure, whereas rest of the 75% is private (of which about 90% is out-of-pocket expenditure). Only 1% of total health expenditure was financed by external resources (WHO 2009). An increase in health spending under a climate change scenario requires that focus be shifted from private to public spending so as to reduce the burden that may fall on households with low incomes or those that

[22] The average for countries in the "low income" group was 4.3 in 2006. The global average was 8.7 (WHO 2009).

are below poverty line, as capacity to cope with climate disasters and their aftereffects is already low in this section of society.

Apart from increasing the total budgetary allocation towards health improvements, the effectiveness of the spending must also be geared up. Although states are responsible for health interventions and spending, the central government is important too, especially in prevention programmes. There is also growing emphasis on devolving health services delivery to elected local governing bodies like the panchayats, as their outreach would be more location specific. Presently, a large disparity can be seen in the health performance and existing health care infrastructure of different states, which creates additional financial challenges. The more backward states,[23] such as Bihar, Madhya Pradesh, Orissa, and Rajasthan, have a high share in total burden of infectious diseases and high rates of infant mortality, child mortality, and child malnutrition (Deolalikar, Jamison, Jha, *et al.* 2008). Such disparity can also be seen between rural and urban areas. Thus, effective health spending would also require identification of most backward and vulnerable states and populations so as to target them with a high coverage of preventive measures; increased access to public health resources (both human and infrastructural); and enhanced technical assistance and awareness creation. Assessing the effectiveness and cost-effectiveness of interventions would also be crucial for preventing imprudent resource allocation.

This study represents several advancements over existing literature on cost assessment of health adaptation. First, the scope of assessment has been extended to include state-level estimates, which can offer interesting insights to policy-makers for assessing potential adaptation actions at more regional and local levels. Estimates of total costs of adaptation at national level provides useful information for international policy debates on adaptation finance, but a downscaling is essential for state and local governments, given the existing diversity (geographical, demographical, and socio-

[23] Also called the Empowered Action Group (EAG) states; constituted in 2001 as an administrative mechanism for the purpose of closely monitoring the implementation of family welfare programmes in poor performing states. Eight states—Uttar Pradesh, Madhya Pradesh, Bihar, Rajasthan, Orissa, Uttarakhand, Chhattisgarh, and Jharkhand—have been identified as EAG states.

economic) in different states of India. Future research is nevertheless needed along these lines, to obtain more robust results based on finer resolution and more comprehensive impact assessment studies.

Second, the study makes an attempt to distinguish between preventive and reactive measures, at least for the malarial burden. As the results show, the cost of prevention is expected to be much higher than the cost of treatment. Little attention has been paid to previous estimates toward identifying prevention separately from case management, and a single methodology is applied to both. This results in considerable underestimation of the cost, as prevention entails a much larger coverage than just treating the infected population. The results for malaria in the present study highlight the fact that prevention will play a crucial role in reducing the mortality burden associated with the disease, and therefore, reduce the need for reactive measures and associated costs that are largely supported by the individuals (private costs) in terms of additional medical expenses, loss of labour hours and productivity, reduced savings, and so on. Another important step related to prevention is the assessment of the population expected to be at risk in the future. This was based on some studies available for India in the malarial context, though extensive research is still needed to model more precisely these impacts in the different geographical areas.

Third, the identification of both preventive and reactive measures included in the cost assessment was based on a review of national programmes and strategies being implemented by the Government in India, complemented by recommendations from the WHO. Indeed, for policy purposes, the cost assessment needs to be as country-specific as possible. The analysis includes a comprehensive set of measures, also taking into account specific population groups that are particularly vulnerable, such as children and pregnant women. The adaptation costs for malaria include costs involved in strengthening the health care system in terms of human resources, training, transport, and office supplies.

Finally, the study takes into account expected future economic development in India, both in terms of incidence projections under development scenarios, and in terms of MDGs targets by incorporating the desired population coverage vis-á-vis protection measures.

ANNEXURES

Annexure I

Projected relative risks for South East Asia Region (SEAR)

SEAR	Relative risks unmitigated emissions		S750			S550			
	Mid	Low	High	Mid	Low	High	Mid	Low	High
Malaria	1.01	1.00	1.02	1.01	1.00	1.01	1.00	1.00	1.01
Diarrhoea	1.09	1.00	1.19	1.07	1.00	1.15	1.06	1.00	1.13
Malnutrition	1.17	1.00	1.33	1.17	1.00	1.35	1.12	1.00	1.25

Annexure II

Major malaria control initiatives implemented over the years in India

Programme	Salient features
National Malaria Control Programme, 1953	Remarkably successful; intensive dichlorodiphenyltrichloroethane (DDT) spraying drive for vector control
National Malaria Eradication Programme, 1958	Focus changes from control to complete eradication of the disease
Urban Malaria Scheme, 1971	Targeted at bringing down the malaria incidence in towns and cities; had been implemented in 131 towns covering a population of 101.1 million, as of 2005
Modified Plan of Operation, 1977	Launched after major resurgence in 1976; primary objective to prevent deaths and reduce morbidity due to malaria in both rural and urban areas with the adoption of selective spraying based on stratifying areas as per the Annual Parasite Incidence
Enhanced Malaria Control Project, 1997–2005	World Bank assisted; implemented in 100 districts of eight states predominantly inhabited by tribal population
National Vector-borne Disease Control Programme, 2008–13	Malaria control operations to be implemented in 93 districts in eight states, in two phases, with overall objective to reduce malaria morbidity by 25% and mortality by 50% over base year 2007.

Annexure III

Some national programmes addressing issues in diarrhoeal prevention and treatment and child health care programmes in India

Programme	Salient features
Central Rural Sanitation Programme, 1986	Objective of improving the quality of life of the rural population and expanding the concept of sanitation to include personal hygiene, home sanitation, safe water, garbage disposal, excreta disposal, and waste water disposal
Accelerated Rural Water Supply Programme, 1972	Extending drinking water coverage to the "not covered" and "partially covered" rural habitations in the country; promote sustainability in the sector; ensure supply of safe drinking water in covered habitations
Integrated Child Survival Programme, 1975	Integrated approach towards converging basic services for improved child care, early stimulation and learning, health and nutrition, water and environmental sanitation for young children, expectant and lactating mothers, other women, and adolescent girls in a community
National Oral Rehydration Therapy Programme, 1985	Promoting use of solution made from widely available oral rehydration salt packets and an alternative home-made solution as a remedy for diarrhoea
Accelerated Urban Water Supply Programme, 1993	Targets safe and adequate water supply facilities to the town and cities in the country; improve the environment and the quality of life in those areas
Child Survival and Safe Motherhood Programme, 1992	Aimed at improving health status and reduce mortality and morbidity in infants, children, and mothers. Also seeks to achieve and sustain the high coverage levels under the Universal Immunization Programme
Total Sanitation Campaign (TSC), 1999	Reduced the incidence of water- and sanitation-related diseases and greatly accelerated the sanitation coverage in rural areas, paying particular attention to awareness creation and health education, especially in schools. Under TSC, states such as Sikkim, Mizoram, Tripura, and Kerala have achieved more than 75% household sanitation coverage
Swajaldhrara Programme, 2002	Promoting community-based rural drinking water supply
The National Rural Water Quality Monitoring and Surveillance Programme, 2004	Responsible for the monitoring and surveillance of all drinking water sources in the country as well as generation of awareness among the rural masses about the water quality issues and the problems related to water-borne diseases
Jalmani initiative, 2008	Installation of simple stand alone water purification systems

Annexure IV

Downscaled state-wise additional annual cost of adaptation, by state, for diarrhoea in India in 2030 ('000 $ at 2004 prices, without discounting)

State	% contribution to total cost	Unmitigated scenario (optimistic–pessimistic)	Mitigation scenario (S550) (optimistic–pessimistic)
Andhra Pradesh	15.6	12 420–47 357	8 949–34 123
Assam	0.8	636–2428	458–1749
Chhattisgarh	1.3	1035–3946	745–2843
Delhi	0.9	716–2732	516–1968
Goa	0.1	79–303	57–218
Gujarat	3.0	2388–9107	1721–6562
Haryana	2.0	1592–6071	1147–4374
Himachal Pradesh	3.1	2468–9410	1778–6780
Jammu Division	2.3	1831–6982	1319–5031
Kashmir Division	1.4	1114–4250	803–3062
Jharkhand	0.6	477–1821	344–1312
Karnataka	6.4	5095–19,428	3671–13,999
Kerala	3.2	2547–9714	1835–6999
Madhya Pradesh	4.9	3 901–14 875	2 810–10 718
Maharashtra	8.8	7 006–26 714	5048–19,249
Manipur	0.1	79–303	57–218
Meghalaya	1.2	955–3642	688–2624
Mizoram	0.2	159–607	114–437
Nagaland	0.1	79–303	57–218
Orissa	4.1	3264–12,446	2352–8968
Punjab	1.6	1273–4857	917–3499
Rajasthan	3.2	2547–9714	1835–6999
Sikkim	0.4	318–1214	229–874
Tamil Nadu	3.8	3 025–11 535	2179–8312
Tripura	1.1	875–3339	631–2406
Uttarakhand	0.8	636–2428	458–1749
Uttar Pradesh	3.6	2 866–10 928	2065–7874
West Bengal	23.9	19 028–72 554	13 710–52 279
India		79 616–303 575	57 367–218 741

Note The total cost mentioned for India is different from the total of the state-wise costs as some states for which data were unavailable were excluded from this downscaling.

References

Aaheim A and Aasen M. 2008. **What do we know about the economics of adaptation?** *CEPS Policy Brief*, No. **150**, 1 February 2008

ADB (Asian Development Bank) and International Food Policy Research Institute. 2009. ***Building Resilience in the Agricultural Sector in Asia and the Pacific.*** Mandaluyong City, Philippines: ADB.

Aggarwal P K and Mall R K. 2002. **Climate change and rice yields in diverse agro-environments of India. II. Effect of uncertainties in scenarios and crop models on impact assessment.** *Climatic Change* **52** (3): 331–343

Ahern M J, Kovats R S, Wilkinson P, Fewand F, Matthies R. 2005. **Global health impacts of floods: epidemiological evidence.** *Epidemiologic Reviews* **27**: 36–45.

Akhtar R, Dutt A, and Wadhwa V. 2002. **Health planning and resurgence of malaria in urban India.** In *Urban Health in the Third World*, pp. 65–92, edited by R Akhtar. New Delhi: AHP Publications

Alberini A and Chiabai A. 2007. **Valutazione degli Impatti dei Cambiamenti Climatici sulla Salute.** Agenzia per la Protezione dell'Ambiente e per i Servizi Tecnici APAT and Centro Euro Mediterraneo per i Cambiamenti Climatici CMCC

Armstrong B, Mangtani P, Fletcher A, Kovats R S, McMichael A J, Pattenden S, Wilkinson P. 2004. **Effect of influenza vaccination on excess deaths occurring during periods of high circulation of influenza: cohort study in elderly people.** *British Medical Journal* **329**: 660–663

Beniston M. 2003. **The 2003 heat wave in Europe: a shape of things to come? An analysis based on Swiss climatological data and model simulations.** *Geophysical Research Letters* **3**: L02202.

Bhattacharya S, Sharma C, Dhiman R C, Mitra A P. 2006. **Climate change and malaria in India**. *Current Science* **90**: 369–375

Black R E, Brown K H, and Becker S. 1984. **Malnutrition is a determining factor in diarrhoeal duration, but not incidence, among young children in a longitudinal study in rural Bangladesh**. *The American Journal of Clinical Nutrition*, **37:** 87–94

Brown K H. 2003. **Diarrhoea and malnutrition.** *Journal of Nutrition* **13**(3): 328S–332S.

Campbell-Lendrum D and Woodruff R. 2007. **Climate change: quantifying the health impact at national and local levels.** In *Environmental Burden of Disease*, edited by A Pruss-Ustun. Geneva: World Health Organization

Caulfield L E, Richard S A, and Black R E. 2004. **Undernutrition as an underlying cause of malaria morbidity and mortality in children less than five years old.** *American Journal of Tropical Medicine and Hygiene* **71**(2): 55–63

Caulfield L E, Richard S A, Rivera J A, Musgrove P, Black R E. 2006. **Stunting, wasting and micronutrient deficiency disorders.** In *Disease Control Priorities in Developing Countries,* 2nd edn, pp. 551–567, edited by D T Jamison, J G Breman, A R Measham, G Alleyne, M Claeson, D B Evans, P Jha, A Mills, and P Musgrove P. New York: The World Bank and Oxford University Press. Details available at <http:www.dcp2.org>, last accessed on 23 December 2009

Census of India. 2001. ***Census Data 2001.*** New Delhi: Office of the Registrar General and Census Commissioner. Details available at <www.censusindia.net>, last accessed on 28 July 2009

Chaudhury S K, Gre J M, and Ray K C S. 2000. **Impact of heat waves in India.** *Current Science* **79**: 153–155

Checkley W, Epstein L D, Gilman R H, Figueroa D, Cama R I, Patz J A, Black R E. 2000. **Effects of El Niño and ambient temperature on hospital admissions for diarrhoeal diseases in Peruvian children.** *Lancet* **355**: 442–450.

Chow J, Darley S, and Laxminarayan R. 2007. ***Cost-effectiveness of Disease Interventions in India.*** Washington: Resources for the Future [Discussion Paper No. 07–53, December 2007]

Christensen J H, Hewitson B, Busuioc, A, Chen A, Gao X, Held I, Jones R, Kolli R K, Kwon W T, Laprise R, Magaña Rueda V, Mearns L, Menéndez C G, Räisänen J, Rinke A, Sarr A, Whetton P. 2007. **Regional climate projections.** In *Climate Change 2007: The Physical Science Basis: Contribution of Working Group I to the Fourth Assessment Report of the Intergovernmental Panel on Climate Change*, edited by S Solomon, D Qin, M Manning, Z Chen, M Marquis, K B Averyt, M Tignor, and H L Miller. Cambridge, UK, and New York: Cambridge University Press

Colwell R R. 1996. **Global climate and infectious diseases: the cholera paradigm**. *Science* **274**(5295): 2025–2031

Confalonieri U, Menne B, Akhtar R, Ebi K L, Hauengue M, Kovats R S, Revich B, Woodward A. 2007. **Human health 2007.** In *Climate Change 2007: Impacts, Adaptation and Vulnerability: Contribution of Working Group II to the Fourth Assessment Report of the Intergovernmental Panel on Climate Change*, edited by M L Parry, O F Canziani, J P Palutikof, P J van der Linden, and C E Hanson. Cambridge: Cambridge University Press

Cruz R V, Harasawa H, Lal M, Wu S, Anokhin Y, Punsalmaa B, Honda Y, Jafari M, Li C, Ninh N H. 2007. **Asia.** In *Climate Change 2007: Impacts, Adaptation and Vulnerability. Contribution of Working Group II to the Fourth*

Assessment Report of the Intergovernmental Panel on Climate Change, pp. 469–506, edited by M L Parry, O F Canziani, J P Palutikof, P J van der Linden, C E Hanson. Cambridge, UK: Cambridge University Press

De U S and Mukhopadhyay R K. 1998. **Severe heat wave over the Indian sub-continent in 1998, in perspective of global climate.** *Current Science* **75**: 1308–1311

Deolalikar A B, Jamison D T, Jha P, Laxminarayan R. 2008. **Financing health improvements in India.** *Health Affairs* **27**(4): 978–990

Dhiman R C, Pahwa S, and Dash A P. 2008. **Climate change and malaria in India: interplay between temperatures and mosquitoes.** *Regional Health Forum* **12**(1): 27–31

Donaldson G C, Kovats R S, Keatinge W R, McMichael A. 2001. **Heat- and cold-related mortality and morbidity and climate change.** In *Health Effects of Climate Change in the UK,* pp. 70–80. London: Department of Health

Edirisinghe J S. 1986. **Infections in the malnourished: with special reference to malaria and malnutrition in the tropics.** *Annals of Tropical Paediatrics* **6**(4): 233–237

Epstein Y and Mills E. 2005. **Climate change futures. Health, ecological and economic dimensions.** The Centre for Health and the Global Environment. Harvard Medical School

Epstein Y, Sohar E, and Shapiro Y. 1995. **Exceptional heatstroke: a preventable condition.** *Israel Journal of Medical Science* **31**: 454–462

EEA (European Environment Agency). 2007. ***Climate Change: the cost of inaction and the cost of adaptation.*** *EEA Technical Report.* [Report no. 13/2007. ISSN 1725–2237]

Faunt J D, Wilkinson T J, Aplin P, Henschke P, Webb M, Penhal R K. 1995. **The effete in the heat: heat-related hospital presentations during a ten-day heat wave.** *Australian and New Zealand Journal of Medicine* **25**: 117–121.

FAO (Food and Agriculture Organization) and World Bank. 2001. ***Farming Systems and Poverty: improving farmer's livelihoods in a changing world***, edited by J Dixon, A Gulliver, and D Gibbon. Rome and Washington, DC: Malcolm Hall

Gaffen D J and Ross R J. 1998. **Increased summertime heat stress in the US.** *Nature* **396**: 529–530

Gawith M J, Downing T E, and Karacostas T S. 1999. **Heatwaves in a changing climate.** In *Climate, Change and Risk*, pp. 279–307, edited by T E Downing, A J Olsthoorn, and R S J Tol. London: Routledge

Gragnolati M, Shekar M, Gupta M D, Bredenkamp C, Lee Y K. 2005. **India's Undernourished Children: a call for reform and action**. [Health Nutrition and Population (HNP), HNP Discussion Paper Series, World Bank, Washington, DC, 2005.]

Gosain A K, Rao S, and Basuray D. 2006. **Climate change impact assessment on hydrology of Indian river basins.** *Current Science* **90**(3): 346–353

GoI (Government of India). 2004. ***India's Initial National Communication to the United Nations Framework Convention on Climate Change.*** New Delhi: Ministry of Environment and Forests, GoI.

Greenough G, McGeehin M A, Bernard S M, Trtanj J, Riad J, Engelberg D. 2001. **The potential impacts of climate variability and change on health impacts of extreme weather events in the United States.** *Environmental Health Perspectives* **109**: 191–198

Guerrant R L, Schorling J B, McAuliffe J F, De Souza M A. 1992. **Diarrhea as a cause and an effect of malnutrition: diarrhoea prevents catch-up growth and malnutrition increases diarrhoea frequency and duration.** *The American Journal of Tropical Medicine and Hygiene* **47**(1): 28–35

Haines A, McMichael A J, and Epstein P R. 2000. **Environment and health: global climate change and health.** *Canadian Medical Association Journal* **163**(6): 729–734

Hajat S, Armstrong B, Gouveia N, Wilkinson P. 2005. **Comparison of mortality displacement of heat related deaths in Delhi, Sau Paulo and London.** *Epidemiology* **16**: 613–620

Hall G V, D'Souza R M, and Kirk M D. 2002. **Foodborne disease in the new millennium: out of the frying pan and into the fire**? *The Medical Journal of Australia* **2**(177): 614–618.

Hijioka Y, Takahashi K, Matsuoka Y, Harasawa H. 2002. **Impact of global warming on waterborne disease**. *Journal of Japan Society on Water Environment* **25**: 647–652

Hogrefe C, Biswas J, Lynn B, Civerolo K, Ku J Y, Rosenthal J, Rosenzweig C, Goldberg R, Kinney P L. 2004. **Stimulating regional-scale ozone climatology over the eastern United States: model evaluation results.** *Atmospheric Environment* **38**: 2627

Hutton G. 2000. **Consideration in evaluating the cost-effectiveness of environmental health interventions.** World Health Organization. [Report no. WHO/SDE/WSH/00.10]

Hutton G and Haller L. 2004. **Evaluation of the costs and benefits of water and sanitation improvements at the global level.** World Health Organization. [Report no. WHO/SDE/WSH/04.04]

IPCC (Intergovernmental Panel on Climate Change). 2001. **Climate change 2001: impacts, adaptation and vulnerability.** In *Contribution of Working Group II to the Third Assessment Report of the Intergovernmental Panel on Climate Change*, edited by J J McCarthy, O F Canziani, N A Leary, D J Dokken, and K S White. Cambridge: Cambridge University Press

IPCC (Intergovernmental Panel on Climate Change). 2007. **Climate change 2007: impacts, adaptation and vulnerability.** In *Contribution of Working Group II to the Fourth Assessment Report of the Intergovernmental Panel on Climate Change*, edited by B Metz, O Davidson, P Bosch, R Dave, and L Meyer. Cambridge: Cambridge University Press

IIPS (International Institute for Population Sciences) and Macro International). 2007. ***National Family Health Survey (NFHS-3), 2005–06: India,*** Vol. 1. Mumbai: IIPS

Jamison D T, Breman J G, Measham A R, Alleyne G, Claeson M, Evans D B, Jha P, Mills A, Musgrove P (eds). 2006. ***Disease Control Priorities in Developing Countries,*** 2nd edn. New York: World Bank and Oxford University Press. Details available at <http://files.dcp2.org/pdf/DCP/DCPFM.pdf>, last accessed on 4 January 2010

Keusch G T, Fontaine F, Bhargava A, Boschi-Pinto C, Bhutta Z A, Gotuzzo E, Rivera J A, Chow J, Shahid-Salles S A, Laxminarayan R. 2006. **Diarrhoeal diseases.** In *Disease Control Priorities in Developing Countries*, 2nd edn, pp. 371–387, edited by D T Jamison, J G Breman, A R Measham, G Alleyne, M Claeson, D B Evans, P Jha, A Mills, P Musgrove. New York: World Bank and Oxford University Press

Kilbourne E M. 1997. **Heatwaves.** In *The Public Health Consequences of Disasters*, pp. 51–61, edited by E Noji. Oxford and New York: Oxford University Press

Kiszewski A, Johns B, Schapria A, Delacollette C, Crowell V, Tan-Torres T, Ameneshewa B, Teklehaimanot A, Nafo-Traore F. 2007. **Estimated global resources needed to attain International Malaria Control goals.** *Bulletin WHO 2007* **85**: 623–630

Kovats R S, Bouma M J, Hajat S, Worrall E, Haines A. 2003. **El Niño and health.** *The Lancet* **362**(9394): 1481–1489

Kovats R S and Ebi K L. 2006. **Heatwaves and public health in Europe.** *European Journal of Public Health* **16**(6): 592–599

Kumar K S. 2007. **Climate change studies in Indian agriculture.** *Indian Economic and Political Weekly.* 17 November 2007: 13–18

Kumar A, Valecha A, Jain T, Dash A P. 2007. **Burden of malaria in India: retrospective and prospective view.** *American Journal of Tropical Medicine and Hygiene* **77** (6): pp. 69–78

Lal M. 2001. **Tropical cyclones in a warmer world.** *Current Science* **80**: 1103–1104

Lal S, Sonal G S, and Phukan P K. 2000. **Status of malaria in India.** *Journal of Indian Academy of Clinical Medicine* **5**(1): 20–23

Langford I H and G Bentham. 1995. **The potential effects of climate change on winter mortality in England and Wales.** *International Journal of Biometeorology* **38**: 141–147

Lipp E K, Huq A, and Colwell R R. 2002. **Effects of global climate on infectious disease: the cholera model.** *Clinical Microbiology Reviews* **15**(4): 757–770

Liu J, Steffen F, van Wesenbeeck C F A, Fuchs M, Liangzhi Y, Obersteiner M, Yang H. 2008. **A spatially explicit assessment of current and future hotspots of hunger in Sub-Saharan Africa in the context of global change.** *Global and Planetary Change* **64**: 222–235

Mall R K, Singh K, Gupta A, Srinivasan G, Rathore L S. 2006. **Impact of climate change on Indian agriculture: a review.** *Climatic Change* **78**: 445–478

Markandya A and Chiabai A. 2009. **Valuing climate change impacts on human health: empirical evidence from the literature.** *International Journal Environmental Research and Public Health* **6**: *759–786*

Martens W J M. 1997. **Climate change, thermal stress and mortality changes.** *Social Science and Medicine* **46**: 331–344

Martens W J M, Jetten T H, Rotmans J, Niessen L W. 1995. **Climate change and vector-borne diseases: a global modelling perspective.** *Global Environmental Change* **5**: 195–209

Mathers C D and Loncar D. 2006. **Projections of global mortality and burden of disease from 2002 to 2030.** *PLoS Med* **3**(11): e442. doi:10.1371/journal.pmed.0030442

McMichael A J and Kovats R S. 2000. **Climate change and climate variability: adaptations to reduce adverse health impacts**. *Environmental Monitoring and Assessment* **61**: 49-64.

McMichael A J, Campbell-Lendrum D H, Corvalan C F, Ebi K L, Githeko A K, Scheraga J D, Woodward A (eds). 2003. ***Climate Change and Human Health: risks and responses.*** Geneva: World Health Organization Press

McMichael A J, Campbell-Lendrum D, Kovats S, Edwards S, Wilkinson P, Wilson T, Nicholls R, Hales S, Tanser F, LeSueur D, Schlesinger M, Andronova N. 2004. **Global climate change.** In *Comparative Quantification of Health Risks: global and regional burden of disease due to selected major risk factors*, pp. 1543–1649, edited by M Ezzati, A Lopez, A Rodgers, C Murray. Geneva: World Health Organization Press

McMichael A A, Githeko R A, Carcavallo R, Gubler D J, Haines A, Kovats R S, Martens P, Patz J, Sasaki A, Ebi K, Focks D, Kalkstein L S, Lindgren E, Lindsay L R, Sturrock R. 2001. **Human health.** In *Climate Change 2001: impacts, adaptation, and vulnerability. Contribution of Working Group II to the Third Assessment Report of the Intergovernmental Panel on Climate Change*, pp. 453–485, edited by J J McCarthy, O F Canziani, N A Leary, D J Dokken, K S White. Cambridge: Cambridge University Press

McMichael A J, Wilkinson P, Sari Kovats R, Pattenden S, Hajat S, Armstrong B, Vajanapoom N, Niciu E M, Mahomed H, Kingkeow C, Kosnik M, O'Neill M S, Romieu I, Ramirez-Aguilar M, Barreto M L, Gouveia N, Nikiforov B. 2008. **International study of temperature, heat and urban mortality: the 'Isothurm' project.** *International Journal of Epidemiology* **37**: 1–11

Menne B and Bertollini R. 2000. **The health impacts of desertification and drought.** *Down to Earth* **14**: 4–6

Menon P, Deolalikar A, and Bhaskar A. 2009. ***The Indian State Hunger Index: comparison of hunger across states.*** Washington, DC, Bonn, and Riverside: IFPRI-Welthungerhilfe-University of California

Ministry of Health and Family Welfare. 2002. ***The National Health Policy 2002***. New Delhi: Ministry of Health and Family Welfare, Government of India. Details available at <www.mohfw.nic.in/NRHM/Documents/National_Health_policy_2002.pdf>, last accessed on 10 January 2010

Ministry of Health and Family Welfare. 2006. ***National Rural Health Mission***. New Delhi: Ministry of Health and Family Welfare, Government of India. Details available at <www.mohfw.nic.in/NRHM/Documents/Mission_Document.pdf>, last accessed on 15 January 2010

Ministry of Health and Family Welfare. 2007. ***The National Drug Policy on Malaria.*** New Delhi: Ministry of Health and Family Welfare, Government of India. Details available at <nvbdcp.gov.in/Doc/Revised%20drug%20policy.pdf>, last accessed on 15 January 2010

MoF (Ministry of Finance). 2008. ***Economic Survey 2007/08.*** New Delhi: Ministry of Finance, Government of India. Details available at <http://indiabudget.nic.in/>, last accessed on 14 January 2009

Mondal N C, Biswas R, and Manna A. 2001. **Risk factors of diarrhoea among flood victims: a controlled epidemiological study.** *Indian Journal of Public Health* **45**: 122–127

Murray C and Lopez A. 1996. ***Global Burden of Disease.*** Boston: Harvard University Press

Nelson G C, Rosegrant M W, Koo J, Robertson R, Sulser T, Zhu T, Ringler C, Msangi S, Palazzo A, Batka M, Magalhaes M, Valmonte-Santos R, Ewing M, Lee D. 2009. ***Climate Change Impact on Agriculture and Costs of Adaptation.*** Washington, DC: International Food Policy Research Institute

Ogden N H, Maarouf A, Barker I K, Bigras-Poulin M, Lindsay L R, Morshed M G, O'Callaghan C J, Ramay F, Waltner-Toews D, Charron D F. 2006. **Climate change and the potential for range expansion of the Lyme disease vector *Ixodes scapularis* in Canada.** *International Journal of Parasitology* **36**: 63–70

Parry M. 2007. **The implications of climate change for crop yields, global food supply and risk of hunger.** *SAT eJournal|ejournal.icrisat.org* **4**(1): 44

Parry M, Rosenzweig C, Iglesiasc A, Livermored M, Fischer G. 2004. **Effects of climate change on global food production under SRES emissions and socio-economic scenarios.** *Global Environmental Change* **14**: 53–67

Pascual M, Rodo X, Ellner S P, Colwell R, Bouma M J. 2000. **Cholera dynamics and El Niño southern oscillation.** *Science* **289**: 1766–1769

Patz J A, Campbell-Lendrum D, Holloway T, Foley J A. 2005. **Impact of regional climate change on human health.** *Nature* **438**(17): 310–317

Piver W T, Ando M, Ye F, Portier C T. 1999. **Temperature and air pollution as risk factors for heat stroke in Tokyo, July–August 1980–1995.** *Environmental Health Perspective* **107**(11): 911–916

Planning Commission. 2008. ***Eleventh Five-year Plan (2007–12), Social Sector,*** Vol. II. New Delhi: Oxford University Press

Purohit S G, Kelkar S D, and Simha V K. 1998. **Time series analysis of patients with rotavirus diarrhoea in Pune, India.** *Journal of Diarrhoeal Diseases Research* **16**(2): 74–83.

Raghavan S and Rajesh S. 2003. **Trends in tropical cyclone impact: a study in Andhra Pradesh, India.** *Bulletin of the American Meteorological Society* **84**: 635–644

Rodó X, Pascual M, Fuchs G, Faruque A S G. 2002. **ENSO and cholera: a nonstationary link related to climate change**? *Proceedings of the National Academy of Sciences of the United States of America* **99**(20): 12901–12906

Rogers D J and Randolph S E. 2006. **Climate change and vector-borne diseases.** *Advances in Parasitology* **62**: 345–381

Rooney C, McMichael A J, Kovats R S, Coleman M. 1998. **Excess mortality in England and Wales, and in Greater London, during the 1995 heat wave.** *Journal of Epidemiology and Community Health* **52**: 482–486

Schultz J M, Russell J, and Espine Z. 2005. **Epidemiology of tropical cyclones: the dynamics of disaster, disease and development.** *Epidemiologic Reviews* **27**: 21–35

Simkhovich B Z, Kleinman M T, and Kloner R A. 2008. **Air pollution and cardiovascular injury: epidemiology, toxicology, and mechanisms.** *Journal of the American College of Cardiology* **52**: 719–726

Singh R B K, Hales S, de Wet N, Raj R, Hearnden M, Weinstein P. 2001. **The influence of climate variation and change on diarrhoeal disease in the Pacific Islands.** *Environmental Health Perspectives* **109**(2): 155–159

Stenberg J, Johns B, Scherpbier R.W, Edeger T T T. 2007. **A financial road map to scaling up essential child health interventions in 75 countries.** *Bulletin WHO 2007* **5**: 305–314

Stevenson D S, Johnson C E, Collins W J, Derwent R G, Edwards J M. 2000. **Future estimates of tropospheric ozone radiative forcing and methane turnover: the impact of climate change.** *Geophysical Research Letters* **27**: 2073–207

Sur D, Dutta P, Nair G B, Bhattacharya S K. 2000. **Severe cholera outbreak following floods in a northern district of West Bengal.** *Indian Journal of Medical Research* **112**: 178–182

Takahashi K, Honda Y, and Emori S . 2007. **Estimation of changes in mortality due to heat stress under changed climate.** *Journal of Risk Research* **10**: 339–354

Tubiello F N and Fischer G, 2007. **Reducing climate change impacts on agriculture: global and regional effects of mitigation, 2000–2080.** *Technological Forecasting and Social Change* **74** (2007): 1030–1056

UN (United Nations). 2003. ***World Fertility Report.*** New York: Population Division, Department of Economic and Social Affairs

UN (United Nations). 2008. ***The Millennium Development Goals Report.*** New York: UN

UNICEF–WHO. 2009. **Diarrhoea: why children are still dying and what can be done.** Details available at <http://whqlibdoc.who.int/publications/2009/9789241598415_eng.pdf>, last accessed on 2 December 2010

UNFCCC (United Nations Framework Convention on Climate Change). 2007. ***Investments and Financial Flows to address Climate Change.*** Bonn: UNFCCC

UNPD (United Nations Population Division). 2003. ***World Population Prospects: the 2002 revisions,*** Vol. 1. ***Comprehensive Tables.*** New York: United Nations. 868 pp.

UNPD (United Nations Population Division). 2005. ***World Population Prospects: the 2004 revisions.*** New York: United Nations. Details available at <http//esa.un.org/>, last accessed on 15 January 2010

van Lieshout M, Kovats R S, Livermore M T J, Martens P. 2004. **Climate change and malaria: analysis of the SRES climate and socio-economic scenarios.** *Global Environmental Change* **14**: 87–99

Van Rensburg J J J and Blignaut J N. 2002. **The economic impact of an increasing health risk due to global warming**, pp. 17–127. [First Conference Proceedings. Forum for Economics and the Environment, Cape Town]

von Grebmer K, Fritschel H, Nestorova B, Olofinbiyi T, Pandya-Lorch R, Yohannes Y. 2008. ***Global Hunger Index: the challenge of hunger.*** Washington, DC and Dublin: Welthungerhilfe-IFPRI-Concern Worldwide.

Watson J T, Gayer M, and Connolly M A. 2007. **Epidemics after natural disasters.** *Emerging Infectious Diseases* **13**(1): 1–5

Wayne P, Foster S, Connolly J, Bazzaz F, Epstein P. 2002. **Wayne production of allergenic pollen by ragweed (*Ambrosia artemisiifolia L.*) is increased in CO_2-enriched atmospheres.** *Annals of Allergy, Asthma and Immunology* **8**: 279–282

WHO (World Health Organization). 2003. ***Climate Change and Human Health: risks and responses. Summary.*** Geneva, Switzerland: WHO

WHO (World Health Organization). 2005. ***The Treatment of Diarrhoea: a manual for physicians and other senior health workers.*** Geneva, Switzerland: WHO

WHO (World Health Organization). 2006. ***Guidelines for the Treatment of Malaria.*** Geneva, Switzerland: WHO

WHO (World Health Organization). 2007. ***Global Malaria Programme. Methodology for estimating the costs of global malaria control (2006–15).*** Geneva, Switzerland: WHO [Technical working paper 2007]

WHO (World Health Organization). 2008a. ***The Global Burden of Disease: 2004 update.*** Geneva, Switzerland: WHO

WHO (World Health Organization). 2008b. ***World Malaria Report, 2008.*** Geneva, Switzerland: WHO

WHO (World Health Organization). 2009. ***World Health Statistics, 2009.*** Geneva, Switzerland: WHO

WHO-RBM-UNICEF-PSI-MSH. 2004. ***Sources and Prices of Selected Products for the Prevention, Diagnosis, and Treatment of Malaria.*** Geneva, Switzerland: WHO.

WHO-UNICEF. 2004. **Clinical management of acute diarrhoea.** Details available at <www.emro.who.int/cah/pdf/who_unicef_statement.pdf>, last accessed on 21 January 2010

Woodruff R E, Hales S, Butler C, McMichael A. 2005. ***Climate Change and Health Impacts in Australia: effects of dramatic CO_2 emission reductions.*** Canberra: Australian National University. 45 pp. [Report for the Australia Conservation Foundation and the Australian Medical Association]

World Bank. 2009. **The costs to developing countries of adapting to climate change: new methods and estimates.** *The Global Report of the Economics of Adaptation to Climate Change Study.* Consultation Draft

Zhou G, Minakawa N, Githeko A K, Yan G. 2004. **Association between climate variability and malaria epidemics in the East African highlands.** *Proceedings of the National Academy of Sciences of the United States of America.* **101**(8): 2375–2380

three

Forest ecosystems

Elena Ojea,[a] Bharat Bhushan Agrawal,[b] Ranjan Kumar Ghosh,[c] and P K Joshi[d]

Background

In general terms, ecosystem resilience is reducing as a consequence of climate change and other anthropocentric causes. The Intergovernmental Panel on Climate Change (IPCC) predicts that approximately, 20%–30% of the known plant and animal species will be subjected to an increased risk of extinction if global temperature increases in the range 1.5–2.5°C (IPCC 2007). In the last report of the IPCC, Fischlin, Midgley, Price, *et al.* (2007) estimate that a 3°C warming would transform about one-fifth of the world's ecosystems. It is also likely that there will be substantial impacts on marine ecosystems with a 3°C warming (UNFCCC 2007a). Biodiversity loss is expected to accelerate in the near future, stressed by climate change (Pimm and Raven 2000; Thomas 2004), and climate change impacts are expected to be the primary cause for biodiversity loss and changes in their services on a global scale by the end of the 21st century (MEA 2005). Biodiversity is a key component of ecosystems, influencing an ecosystem's capacity to respond to external stresses such as

[a] Basque Centre for Climate Change (BC3), Bilbao, Spain.
[b] The Energy and Resources Institute (TERI), New Delhi, India.
[c] Division of Resource Economics, Humboldt University Berlin, Berlin, Germany, and The Energy and Resources Institute (TERI), New Delhi, India.
[d] Associate Professor, TERI University, New Delhi, India.

climatic impacts. Moreover, biodiversity supports ecosystem capacity to regulate climate. Maintaining and enhancing the resilience of natural ecosystems have been advocated as the best and most cost-effective defence against climate change (Zaunberger, Stefan, and Ladislav 2008).

When assessing the impacts of climate change on biodiversity and ecosystems, there are several key issues that may be taken into account. First, there is an increasing evidence on the physical and biological impacts of regional climate changes over time, as reported by the IPCC's *Fourth Assessment Report (AR4)* (IPCC 2007). Second, global losses of biodiversity are usually irreversible and, therefore, of key relevance. Third, climatic impacts on biodiversity and ecosystems are difficult to be isolated from other anthropocentric causes such as land-use change, pollution, and overexploitation of resources. Fourth, the capacity of the ecosystems to adapt to external stresses (resilience) is decreasing due to climate change. All these factors point towards the urgent need for planning adaptive actions in biodiversity and ecosystems, as well as addressing the difficulties that may arise when computing costing for these adaptation actions.

This chapter aims at developing a methodological framework for estimating the costs of adaptation for ecosystems and biodiversity. A detailed methodology has been developed based on the existing literature on adaptation costs by including a set of possible adaptation options, identifying areas that are more vulnerable to climate change, and identifying the magnitude of the adaptation option needed, including both positive and negative impacts. This methodology provides a general framework that can also be applied to other ecosystem types. In this chapter, the focus is on Indian forests so to estimate the costs of adaptation due to the expected climate change impacts.

General impacts of climate change on biodiversity and ecosystems

According to the IPCC's *AR4*, there is strong evidence of climate change already affecting biodiversity and ecosystems in the Northern Hemisphere (IPCC 2007). Terrestrial ecosystems will suffer from a decrease in their resilience capacity due to an unprecedented combination of climate change, associated disturbances such as flooding or wildfires, and other global change drivers like overexploitation of resources (IPCC 2007). At the same time, biodiversity is showing consistent responses to climate change (IPCC 2007). The main climate change impacts on biodiversity and ecosystems, as observed in the Northern Hemisphere and compiled by the IPCC, are presented in Table 1.

Table 1 Main observed impacts of climate change on biodiversity and ecosystems as observed in the Northern Hemisphere

Impacts affecting	Type	Impacts already observed
Ecosystems	Lake and river ecosystems	▪ Abundance and productivity effects on freshwater species (eutrophication) ▪ Effects in community composition of freshwater species ▪ Effects on migration and distribution of freshwater species
	Forest	▪ Significant lengthening/shortening of the growing season (phenological variations) ▪ Increase in forest productivity in some regions ▪ Drought reducing forest productivity ▪ Increase in forest fires ▪ Migration and distribution of the species
Biodiversity	Terrestrial species	▪ Earlier onset of spring events ▪ Increased net primary production linked to longer growing season ▪ Migration and lengthening of the growing season ▪ Changes in abundance of certain species ▪ Changes in community composition
	Marine and freshwater species	Pole-ward shifts in ranges and changes in algal, plankton, and fish abundance in high-latitude oceans

Source IPCC (2007)

Current responses to climate change manifest the importance of global change and serve as a basis for studying and predicting future expected impacts. Fischlin, Midgley, Price, *et al.* (2007) identify ecosystems based on the vulnerability to climate change, wherein the most vulnerable ecosystems are those that are expected to experience the most severe ecological impacts, including species extinctions and major biome changes. These ecosystems are tundra, boreal forests, mountains, and Mediterranean-type ecosystems on continents; mangroves and salt marshes along coasts; and coral reefs and sea-ice biomes in oceans. On the other hand, the least vulnerable ecosystems are those that are expected to suffer a positive initial ecological impact, such as increased net primary productivity (NPP), and include savannas and species-poor deserts. The IPCC has projected a series of impacts on biodiversity and ecosystems for expected changes in physical conditions and extreme events. Table 2 presents a summary of those impacts for forest ecosystems and biodiversity. Impacts are presented in relation to the main changes in the climate conditions.

Table 2 Impacts on forests and biodiversity, as projected by the IPCC

Change in conditions	Impacts on forests	Impacts on biodiversity
Changes in structure and functioning of terrestrial and marine ecosystems, with a global warming of 2–3 °C[a]	▪ Increase of 2 °C, 15% ecosystems affected; increase of 3.5 °C, 40% ecosystems affected ▪ Disturbance regimes: wildfire and insects ▪ Major biome changes/shifts, including emergence of novel biomes	▪ Increasing high risk of extinction for roughly 20%–30% (of species assessed so far varying among regional biotas from 1% to 80%) ▪ Changes in species–ecology interactions, with predominantly negative consequences for goods and services ▪ Shifts in species distribution
Ecosystems are very likely to be exposed to atmospheric CO_2 levels much higher than those in the past 650 000 years	▪ Accelerated release of carbon from vulnerable carbon stocks (peatlands, tundra frozen loess, permafrost soils, and soils of boreal and tropical forests) ▪ Carbon emission increase due to albedo changes, wildfire, forest declines at taiga's equatorial limit and methane losses from tundra will be greater than the increase in carbon sequestration by taiga ▪ Tropical forest sequestration by 2100 is likely to be dominated by climate change impacts, especially in drier regions	

Continued...

Table 2 *Contd...*

Change in conditions	Impacts on forests	Impacts on biodiversity
The resilience of many ecosystems is likely to be exceeded by 2100	Key ecosystem properties (like biodiversity) or regulating services (like carbon sequestration) are very likely to be impaired	▪ Extractive use from and fragmentation of wild habitats are very likely to impair species adaptation ▪ Disruption of species-ecology interactions
Forestry production will be affected by the projected changes in the frequency and severity of extreme climate events	▪ Food and forestry trade is projected to increase in response to climate change, with increased food-import dependence of most developing countries ▪ Increase in the incidence of wildfires ▪ Production increase will shift from low-latitude regions in the short term to high-latitude regions in the long term	

[a]Global mean annual temperature change relative to 1880–1999 (°C).

Sources IPCC (2007); Easterling, Aggarwal, Batima, *et al.* (2007)

Climate change impacts on forest ecosystems in India

Developing countries are expected to face the greatest and earliest burden in terms of adverse impacts on livelihoods, economic growth, and welfare (UNDP 2007). Current impacts and predicted overall effects of climate change on biodiversity and ecosystems, summarized in the previous section, are also relevant for the Asian continent, including India. Observed changes in terrestrial ecosystems have become more pronounced in Asia during the last decade. According to the IPCC, climate change is leading to land-use change and population pressure in most parts of Asia. Habitat fragmentation, together with climate change, is expected to increase the risk of extinction of many flora and fauna species. Also the threats to the ecological stability of wetlands, mangroves, and coral reefs around Asia would increase. Forest expansion will be limited by an increase in forest fires (IPCC 2007).

In South Asian countries, climate change will intensify other environmental problems such as exploitation of natural resources (IPCC 2007). Vulnerability to climate change increases because of these non-climate stresses, as the resilience of the ecosystems decreases, affecting exposure and sensitivity as well as adaptation capacity. Less developed regions that are characterized by poverty and unequal access to resources are more affected. Also, the loss of biodiversity has implications for poverty reduction and the well-being of women. As an illustration, deforestation in Orissa, India, has severely affected the role of women as forest produce gatherers (EEA 2008). Women have to walk four times the distance to collect forest products, affecting their livelihood prospects. This loss reduces income, increases drudgery, and affects health. The same report shows that the relative status of women is higher in well-forested villages, and their contribution to the household income is greater than the women in villages that lack this natural resource.

The IPCC has predicted a series of impacts on biodiversity and ecosystems for expected changes in physical conditions and extreme events. Regional predictions for projected impacts in South Asia are summarized in Table 3.

Table 3 Impacts on biodiversity and ecosystems for South Asia, as projected by the IPCC

Change in conditions	Impacts on South Asian forests
▪ Increase in temperature of 0.68 °C per century, increasing trends in annual mean temperature, warming more pronounced during post-monsoon and winter ▪ Increased vulnerability to impacts on food and fibre, biodiversity, coastal ecosystems, and land degradation	▪ Increment on the extinction rate of species due to synergistic effects of climate change and habitat fragmentation. Shifts of terrestrial species to higher elevations ▪ Loss of much of the original forests and biodiversity by 2100 due to the intensification of forest utilization (increase the rate of deforestation). Increase in frequency and intensity of forest fires and pest outbreaks

Sources IPCC (2007); Fischlin, Midgley, Price, *et al.* (2007); Easterling, Aggarwal, Batima, *et al.* (2007); Cruz, Harasawa, Lal, *et al.* (2007)

The forests in South Asia are rich in biodiversity. However, biodiversity is under threat in these areas because of land degradation and resource overexploitation. About 10% of the world's high-altitude plant and animal species occur in the Himalayas; however, this biodiversity is seriously threatened, with 10% of the species listed in 1995 facing such threat (Ravindranath and Murthy 2003). These authors state that the most likely climate scenarios will lead to shifts in forest types, with serious implications for biodiversity and timber production. Additionally, nearly 200 000 villages are heavily dependent on forest resources for their livelihoods in India. Climate change is predicted to impact forests by changing the biome types, which will impact the supply of forest products both to industry and those communities that are heavily dependent on forest goods and services (DEFRA 2005).

Some studies analyse climate change impacts on Indian forest ecosystems. Ravindranath, Sukumar, and Deshingkar (1997) use a model relating climate variables (mean temperature and precipitation) with vegetation types to show how climate change is leading to shifts in the area of different forest types in the Western Ghats. Under the most likely climate scenario, a shift towards wetter types of vegetation is projected for 2020 and 2050 for Uttara Kannada in the Western Ghats, resulting in the loss of area of the drier forest types. Under the worst case scenario, the area under drier forest types of Uttara Kannada will increase at the expense of area under moister forest types, by 2050.

More recently, Ravindranath, Joshi, Sukumar, *et al.* (2006) studied the expected impacts of climate change on forest ecosystems for the entire India, in the long term. The main conclusion drawn is that under the climate projections of 2085, 77% and 68% of the forested grids in India are likely to experience a shift in forest types, under A2 and B2 scenarios, respectively. The authors also identify a likely shift towards wetter forest types in the north-eastern region and towards drier forest types in the north-western region in the absence of human influence. An increase in CO_2 emissions will cause a doubling of the net primary productivity (NPP) under A2 scenario and almost a 70% increase in the B2 scenario. Under the expected shifts in forest distribution, biodiversity is expected to suffer, especially in the transition areas between forest types. However, the overall NPP is likely to increase significantly, mostly due to the carbon fertilization effect (DEFRA 2005).

The need for adaptation

Adaptation becomes necessary because climate change impacts on ecosystems cannot be entirely avoided in the next decades by any mitigation strategy, although mitigation is still necessary in order to reduce future costs of adaptation measures and longer term impacts (IPCC 2007). In an optimistic scenario, mitigation will start protecting people and ecosystems from climate change from 2030 onwards (UNFCCC 2007b). For ecosystems, planned adaptation is necessary to avoid the expected impacts due to climate change, as autonomous adaptation is inadequate (Fischlin, Midgley, Price, *et al.* 2007; Parry, Arnell, Berry, *et al.* 2009). When defining the adaptation strategies concerning ecosystems and biodiversity, the main actions proposed by the IPCC are those that are related to the reduction in other threats, such as habitat degradation, pollution or introduction of alien species. There are currently many international multilateral environmental agreements (MEAs) signed by numerous countries that are already working towards addressing these kinds of threats, such as the Convention on Biological Diversity (CBD), the Ramsar Convention, and the Millennium Ecosystem Assessment, among others. Adaptation measures for biodiversity should be integrated, as well as synergistic responses in adherence to and compliance with the terms and conditions of MEAs to improve environmental quality should also be factored in (IPCC 2007). When designing adaptation measures, it is important to bear in mind the promotion of win-win strategies and the avoidance of maladaptation or unsustainable adaptation. There are adaptation measures that can be beneficial to one sector but can cause significant damages to other sectors. A review on adaptation and mitigation options in relation to the potential impacts on biodiversity is provided in Paterson, Araujo, Berry, *et al.* (2008). In contrast, other adaptation measures can have co-benefits in terms of climate resilience.

In the Stern Review, it is stated that climate change will require nature conservation efforts that extend the current approach of fixed protected areas (Stern 2006). Stern proposes that conservation efforts will be required at the landscape level, with larger areas managed to better accommodate species movement. But ecosystem adaptation options should not focus only on protected areas. Heller and Zavaleta (2009) have reviewed the literature addressing biodiversity management and adaptation in the face of climate change. From 113 papers, they identified 524 recommendations that are ranked based on the frequency of appearance. The general conclusions from the analysis are that 33% of the recommendations address biodiversity conservation in conjunction with ecosystem services. The first three measures ranked are: (1) the increase in habitat connectivity, (2) the integration of climate change into planning exercises, and (3) the mitigation of other threats to ecosystems (Heller and Zavaleta 2009). Based on the literature survey, Tables 3 and 4 list out the main adaptation options proposed per impact, both for biodiversity and forest ecosystems. Forest ecosystems were chosen, as more information on impacts and adaptation is available for forests than for other ecosystems. Furthermore, in the present chapter, a proposed methodology for estimating costs is presented for Indian forests.

Table 4 Adaptation measures for forest ecosystems

Impact	Adaptation activity	Specific policies
Changes in forest productivity	▪ Modification of the rotation times	▪ Flexibility regarding silvicultural management
	▪ Replanting	▪ Replanting with different species ▪ Replanting with populations from other parts of the distributional range (for example, from lower elevations or latitudes) ▪ Schemes to establish fuelwood orchards. Establishment of forest plantations, reducing impacts on forest ecosystems due to increased fuelwood demand
	▪ Alternation of thinning procedures	▪ Flexible policies for salvaging harvests so as to allow adjustments as climatic conditions change
Increase of wildfires and pest outbreaks	▪ The reduction or prevention of fires	▪ Flexible policies for fire control
	▪ Responsible use of herbicides, pesticides, and fertilizers to avoid insect outbreaks and diseases	
	▪ Breeding of pest-resistant lines	

Continued...

Table 4 *Contd...*

Impact	Adaptation activity	Specific policies
Loss of forest resilience	▪ Connectivity, heterogeneity, and diversity at the landscape level, facilitating forest resilience	▪ Forestry schemes with high planting densities ▪ Schemes to plant tree species drawn from warmer climate zones ▪ Policies to include a mix of planting practices ▪ Planting of trees that have greater resistance to heat and drought on the southern range of managed forest boundaries if trees can survive in the current climate ▪ Schemes to plant tree species and adjust management practices considering future climate scenarios ▪ Schemes for forest rehabilitation through agroforestry ▪ A mix of different timber harvesting strategies to promote forest diversity
	▪ Conservation of forest diversity	▪ Schemes to increase forest seed banks. Enhance forest seed banks aimed at conservation of original genetic diversity of forests, to rebreeding. Seed collections should represent the variety of genotypes that exist for each species.
Shifts in forest distribution	▪ Assist or enable ongoing natural adaptive processes	▪ Schemes to help species dispersal and migration ▪ Schemes to enable population mortality and colonization ▪ Schemes to allow changes in community composition or species dominance and changing disturbance regimes
	▪ Reduction of habitat fragmentation	▪ Incentive policies and programmes for multiple-use management that balances preservation and use within a single parcel ▪ Negotiation of conservation easements that protect geographically important land parcels from development
	▪ Development of migration corridors	

Sources FAO (2008); GoI (2008); Heller and Zavaleta (2009)

Previous literature

In the previous section, we have summarized the main climatic impacts on ecosystems and biodiversity and given a detailed account of a broad range of adaptation options for addressing these impacts. In spite of the imminent impacts expected, countries still lack estimates of the costs they need to incur to adapt to climate change. This gap is larger for adaptation in ecosystems where current estimates on cost of adaptation are scarce and global. Moreover, although many actions are needed to adapt to climate impacts, current estimates of adaptation costs are often based on one single adaptation measure such as increasing protected areas (James, Gaston, and Balmford 2001; Balmford, Bruner, Cooper, *et al.* 2002).

The IPCC states that adaptation strategies can often be implemented at low cost; however, comprehensive estimates of adaptation costs and benefits are currently lacking (IPCC 2007). The literature on adaptation costs and benefits is limited and fragmented and is not focused on biodiversity or ecosystems (IPCC 2007; Stern 2006). Moreover, the emphasis is on the US and other Organisation for Economic Co-operation and Development (OECD) countries, with only a few studies for developing countries. Improving the knowledge of the costs of climate change adaptation will allow policy-makers to consider optimal strategies for implementing adaptation policies (IPCC 2007). Estimates of the costs of adaptation for ecosystems are scarce and very recent. Two types of approaches have been developed to estimate adaptation costs: financial flows on conservation and costs of specific adaptation measures.

Financial flows

The United Nations Framework Convention on Climate Change (UNFCCC) has presented an overview of current investment and financial flows by source of financing. They observe that between 1991 and 2000, the Global Environmental Facility (GEF) provided about $1.1 billion in grants and leveraged an additional $2.5 billion in co-financing biodiversity-related projects. Most of these grants were used to support protected areas covering 226 million hectares (Mha) of area in 86 countries (UNFCCC 2007b). James, Gaston, and Balmford (2001) report that in the mid-1990s, an average of

$6.8 billion per year was spent on global protected areas, with about 89% of that amount spent in developed countries. The study of James, Gaston, and Balmford is an attempt to estimate the investment and financial flows needed to protect natural ecosystems from current threats. Towards this end, the authors follow the International Union for the Conservation of Nature (IUCN) recommendations of how much additional land needs to be set aside as biodiversity protection areas, not specifying the additional protection needs that climate change might require. As a result, they estimate that this improved protection will be achieved with an annual increase in expenditures of the order of $12–22 billion (James, Gaston, and Balmford 2001). More recently, Berry (2007) has differentiated between two different approaches for assessing the costs for adaptation: the costs for maintaining natural ecosystems and their services in the face of climate change and the additional costs for the additional adaptation actions. They provide a guide to the costs of adaptation in natural ecosystems, adopting the following methodology: (1) estimating current global expenditures on conservation; (2) estimating the shortfall in current conservation expenditure; and (3) estimating levels of additional expenditure needed for climate change adaptation. Finally, by adopting the estimates of James, Gaston, and Balmford (2001), Berry (2007) obtained the additional conservation costs for an increase in 10% of global protected areas for two scenarios: all protected areas and only highly protected areas. With the assumption that expenses in protected areas are one-third of total biodiversity expenses, they calculated total expenses for biodiversity conservation, which amount to $64.5 billion per year under the scenario with mitigation and $36 billion per year under the scenario with adaptation. Parry, Arnell, Berry, *et al.* (2009) have summarized in a recent report the efforts taken to compute the cost of adaptation worldwide and concluded that the estimates by the UNFCCC (Berry 2007), based on bringing the current protected area cover up to an arbitrary level, are the best approximation available for ecosystems adaptation.

Table 5 summarizes the studies carried out to date with global financial investment on adaptation for biodiversity conservation. We include information about the adaptation measure, the scope of the study, the methodology used, and the estimated costs. These costs have been calculated on a range starting from $12 billion a year and ending at $28 billion a year when only considering the expansion of protected areas as the adaptation measure, and between $65 and $83.5 billion a year when considering expansion of protected areas as one-third of total adaptation measures required.

Table 5 Financial flows in conservation

Study	Adaptation measure	Area covered	Methodology	Costs
James, Gaston, and Balmford (2001)	Increase in PAs	Global	Costs of a 10% increase in all PAs globally as recommended by the IUCN	$12 billion a year
James, Gaston, and Balmford (2001)	Increase in highly PAs	Global	Costs of a 10% increase only in highly PAs	$21.5 billion a year
Balmford, Bruner, Cooper, *et al*. (2002)	Increase in PAs	Global	Cost of expanding PAs to cover 15% of the area on each terrestrial region	$20–28 billion a year
Balmford, Bruner, Cooper, *et al*. (2002)	Increase in marine PAs	Global	Cost of expanding PAs to cover 30% of marine areas	$23 billion a year
Berry (2007)	Adaptation needed for biodiversity conservation in terrestrial ecosystems	Global	Employs James, Gaston, and Balmford (2001) scenarios and estimated costs of a 10% increase in PAs and assumes that expenses in PAs are one-third of total biodiversity expenses	$83.5 billion a year
Berry (2007)	Adaptation needed for biodiversity conservation in marine ecosystems	Global	Employs James, Gaston, and Balmford (2001) scenarios and estimated costs of a 10% increase in PAs and assumes that expenses in PAs are one-third of total biodiversity expenses	$64.5 billion a year

Continued...

Table 5 *Contd...*

Study	Adaptation measure	Area covered	Methodology	Costs
Berry (2007)	Adaptation needed for biodiversity conservation in terrestrial ecosystems	Global	Employs James, Gaston, and Balmford (2001) scenarios and estimated costs of a 10% increase in highly PAs and assumes that expenses in PAs are one-third of total biodiversity expenses	$65 billion a year
Berry (2007)	Adaptation needed for biodiversity conservation in marine ecosystems	Global	Employs James, Gaston, and Balmford (2001) scenarios and estimated costs of a 10% increase in highly PAs and assumes that expenses in PAs are one-third of total biodiversity expenses	$36 billion a year

IUCN – International Union for the Conservation of Nature; PAs – protected areas

Sources Adapted from Berry (2007) and Parry, Arnell, Berry, *et al.* (2009)

Costs of specific adaptation measures

Apart from the global estimates of adaptation costs, a few other regional studies calculating the cost of specific adaptation measures are available in the literature. These studies cover specific adaptation actions such as setting up migration corridors in Kenya (Ferraro and Kiss 2002), coastal reforestation in Croatia (Pagiola, von Ritter, and Bishop 2004) or conservation of tropical forests in Costa Rica (Ferraro and Kiss 2002). From the review of the literature, it is evident that there is a need for a methodology for computing the costing of ecosystem adaptation that takes into account more details, such as (1) identifying where adaptation actions are needed (vulnerable regions), (2) identifying the positive or negative directions of the impacts on each area, (3) linking adaptation options to specific impacts, and (4) valuing the costs of these specific and feasible adaptation requirements. As far as we are concerned, none of these studies attempts to estimate total adaptation costs on a developing country level, taking into account different adaptation measures and their likelihood of being implemented. Moreover, positive adaptation measures such as allowing ecological changes in some cases have not yet been considered in the literature. The contribution of the present work is on presenting the requirements and steps to be followed for reaching such a detailed approach, and testing this methodology for the expected shifts in forest distribution in India.

Methodology

This section describes the proposed methodology for costing adaptation in ecosystems. Based on the literature review and on the identified gaps in knowledge, we propose a methodological guide to estimate the costs of adaptation for ecosystems at the country level. This methodology is potentially applicable to many different ecosystems, although in the present chapter, the approach will be illustrated with forest ecosystems.

Methodology for estimating adaptation costs in ecosystems

The approach is divided into five steps: (1) the identification of the climatic impact; (2) based on the impacts, the identification of the vulnerable areas; (3) the identification and selection of the adaptation options; (4) estimating the unit costs of the adaptation measures; and (5) estimating the total cost of the adaptation measure. Each step is described below, while next section will present how this approach has been followed for forests in India.

Quantification of direct climate change impacts

The first step is to identify and quantify the relevant impacts of climate change on the country's ecosystems. Impacts should be quantified based on the available scientific knowledge, where geographically disaggregated data will be preferred. Most likely, the unit will be the landscape unit or the hectares affected by a specific climate change impact (Tables 2 and 3). Whenever possible, having these impacts on a geographical basis (using a geographical information system [GIS]) will enhance the approach.

Identification of the vulnerable areas

Climate change is not affecting all regions in the same way. It can be the case of a net global impact being neutral at the state level, while the same impact can be extremely negative for a given region or area. In order to avoid this problem, we need to identify the regions in which an impact has significant effects, positive or negative. These areas will be selected for each specific impact identified in the previous step.

Identification of adaptation options

The third step relates to the identification of the relevant adaptation options per impact and vulnerable area. Adaptation measures should be selected based on the criteria listed below.

- *Relevance* The adaptation option should be relevant for the impact. This step can be based on the existing literature on adaptation policies and on the country-specific existing policies.
- *Effectiveness* This is a measure of the success of the adaptation measure on avoiding the impact, that is, how much adaptation measure units are required to avoid the impact to occur or reduce it to the minimum. With the magnitude of the impact identified, we need a rule that gives us the magnitude of the adaptation option to be implemented. If the adaptation option recovers the system to the *status quo* scenario (with no impact), then the efficiency of the adaptation measure will be 1. This measure of efficiency will be employed only when relevant information is available or some imputation can be made reasonably.
- *Scale of action* This refers to the magnitude of the adaptation option. We must have an estimate of the magnitude of the impact to be avoided by the adaptation option in order to be able to calculate the magnitude of the adaptation measure needed, following the effectiveness criteria.
- *Feasibility* This is the real possibility of applying the adaptation option. This is a crucial criterion, wherein both the adequacy to implement the adaptation measures in the vulnerable areas and the feasibility of the adaptation measure must be examined.

Imagine we have a situation where there are impacts (I_i) affecting ecosystems in vulnerable areas. Our baseline scenario is the current situation (current impact) that we identify as I_0. In a medium timeframe and with no adaptation, climate change is projected to change the magnitude of the impact to I_1. The magnitude of an impact will be the difference between the projected impact (I_1) and the baseline scenario (I_0) for all the vulnerable areas (*va*). Thus, an impact *i* can be quantified as

$$I_i = \sum_{va} (I_{1va} - I_{0va})$$

Now for each impact (*i*), we have identified several adaptation measures (*a*). Each adaptation measure has a different effectiveness in reducing the impact, leading to a reduction in the magnitude of the impact. Thus, with the help of an adaptation option, the expected impact will change from I_1 to I_2, where $I_1 > I_2$.

The effectiveness of a given adaptation option (E_a) is the part of the expected impact with climate change (I_1) that can be avoided by the adaptation option $I_1 - I_2$.

$$E_a = \frac{(I_1 - I_2)}{I_1}$$

When the adaptation option completely avoids the impact of climate change (I_1), the effectiveness is equal to 1 and the expected impact with adaptation is equal to zero.

$$\frac{(I_1 - I_2)}{I_1} = 1; I_2 = 0$$

For each impact, we should select the set of adaptation measures that minimize the expected impact with climate change. However, the effectiveness of adaptation options is not always known, and in the example presented later in this chapter, we will deal with this situation.

Identification of per unit cost of the adaptation measure

Based on the existing literature and national expenses on conservation, afforestation, and other measures, the per unit costs must be identified for each adaptation measure. Per unit costs must be in the form of cost per magnitude per year, where the magnitude should be in the same units as the adaptation option. Naidoo and Ricketts (2006) differentiate several conservation costs that can be identified when measuring adaptation; these are, acquisition costs (costs of acquiring property rights of a land area); management costs (costs associated with the conservation programmes developed in a protected area); damage costs (associated with damages to economic activities arising from conservation programmes); and transaction costs (costs of transfer of property rights) and opportunity costs; (costs of foregone opportunities). Per unit costs of the adaptation measure should optimally account for acquisition and management costs.

Total costs

The final step relates to the aggregation of the costs of adaptation measures for all the vulnerable areas and adaptation measures passing the four criteria. For this, first, calculate the cost of each adaptation measure using the unit cost of the adaptation measure and the magnitude of adaptation measure required; second, aggregate the adaptation costs per impact; and third, aggregate the cost of all the impacts to obtain the total cost of adaptation for a given ecosystem.

Although difficult to quantify, indirect impacts and interactions with other sectors, such as agriculture, water, and livelihoods, should be taken into consideration. As it is difficult to account for these interactions at this point, we are not including these indirect impacts into the methodology in a quantitative way. However, interactions with other sectors may be important. One example is the deforestation in Orissa (India), which has severely affected the role of women as forest gatherers with implications on their livelihoods (EEA 2008).

Data

Hereafter, we apply this methodology to forests in India and present estimates for the costs of adaptation. But before this, this section presents the data employed for quantifying the climate change impacts on Indian forests, as well as how Indian forests can be classified in terms of the main biomes and the biodiversity associated with them.

The vegetation model

The impact under consideration is the expected shift in forest distribution in India. The quantification of the impact is obtained from a study by Ravindranath, Joshi, Sukumar, *et al.* (2006), assessing the impact of climate change on Indian forests. The analysis is based on the spatial distribution of forests in India as obtained from the BIOME4 model, which uses a high-resolution regional climate model (HadRM3) of the Hadley Centre as input for climate projections.

The BIOME4 model uses climate-related input variables such as the monthly mean temperature (in °C), precipitation (in mm), monthly sunshine hours (% of maximum), and also data on soil and water holding capacity. Projected values of climate variables for the future scenarios A2 and B2 for 2085 were given by the Indian Institute of Tropical Meteorology (IITM), which had been obtained from the runs of the HadRM3 adapted for the Indian region (Ravindranath, Joshi, Sukumar, *et al.* 2006). The main conclusion is that, under the climate projection for 2085, 77% and 68% of the forested grids in India are likely to experience a shift in forest types under A2 and B2 scenarios, respectively. As a conclusion, the authors obtain a shift towards wetter forest types in the North East region and drier forest types in the north-western region in the absence of human influence.

Forest types in India

India has a geographical area of 328 Mha, of which 64 Mha is covered with forests (Ravindranath 1998). Table 6 shows the acronyms for the forest cover types used in this chapter. Tropical xerophytic shrubland (TPXS) and tropical savannah (TPS) are spread mainly over semi-arid regions like central India. These regions bear heavy human and biotic pressure in the form of grazing, lopping, illegal trade of forest products, intentional fires, and removal of timber for building house, to name a few (Negi 1993). Tropical deciduous forests/woodlands (TPD/WL) are found in both northern and southern parts of India. Most of these forests have high economic value. The fauna is very similar to that found in the previous vegetation type; however, these regions are habitat for some rare and threatened species like the Asiatic lion (*Panthera leo*), black buck, and chinkara (Singh, Singh, and Gupta 2006). Tropical evergreen (TPEG) and warm mixed (WM) forests are generally found in the eastern Himalayas (north-eastern states), parts of West

Bengal, and in Uttarakhand Himalayas, at heights 1500–3000 m. The tropical semi/moist deciduous (TPSD) vegetation type is spread around the Andaman and Nicobar Islands, Himalayan foothills, eastern parts of Western Ghats, Chotta Nagpur belt, and some regions of north-western India, and holds a high level of endemism. Tropical evergreen forests (TPEG) are found mostly in the Western Ghats, upper reaches of Assam and Arunachal Pradesh in the North East, some parts of Andaman and Nicobar Islands, Tamil Nadu, and Himachal Pradesh. Montane evergreen forests (ET/M) are mainly found in the upper reaches of the Nilgiris, Anamalai, and Tirunelveli hills in South India. Remaining important vegetation types are temperate conifer forest (TMC), temperate sclerophyll woodland (TMSW), cool conifer forest (CC), and cold mixed forest (CLDMX), which are all spread in different parts of the Himalayan ranges. They are also a very important biodiversity hotspot, extremely rich in terms of natural resources.

Table 6 Forest vegetation cover types in India

Vegetation types	
TPXS	Tropical xerophytic shrubland
TPD/WL	Tropical deciduous forest/woodland
WM	Warm mixed forest/temperate evergreen/broadleaved
TPSD	Tropical semi-deciduous forest
TPS	Tropical savanna
TPEG	Tropical evergreen forest
TMC	Temperate conifer forest
TMSW	Temperate sclerophyll woodland
CC	Cool conifer forest
ET/M	Evergreen taiga/montane forest
CLDMX	Cold mixed forest

Source Ravindranath, Joshi, Sukumar, *et al.* (2006)

Results

In this section, a detailed description of how we have implemented the methodology for costing ecosystem adaptation to Indian forests is provided. We follow each step of the methodological approach previously developed, and obtain an estimation of the costs of adaptation for forests in India. This methodology allows us to come closer to the identification and scale of adaptation options relevant to the specific impacts on a regional basis. From there, we identify the costs of applying those options and, in the end, aggregate them to obtain the total costs.

Quantification of direct effects of climate change

The main impact affecting forest ecosystems in India is the expected shift between forest types (Joshi, Roy, Singh, *et al.* 2006; Ravindranath, Joshi, Sukumar, *et al.* 2006). This shift will affect forests in terms of their productivity and dominant species, as well as the biodiversity and human livelihoods dependent on each forest type.

The impact quantification is restricted to the extent of sound scientific information that is currently available. We have followed the study of Ravindranath, Joshi, Sukumar, *et al.* (2006), which allows us to identify the forest types that will be impacted the most by climate change in terms of change in their distribution. The original study analysed 35 190 forested grids in India and showed that 77% of grids under the A2 scenario and 68% under the B2 scenario are likely to undergo vegetation change by 2085. For our analysis, we consider only the B2 scenario, as the detailed matrix for the distribution of shifts between forest types is not available for the A2 scenario.

Table 7 presents the magnitude of the expected shifts between forest types, as obtained from Ravindranath, Joshi, Sukumar, *et al.* (2006), under the B2 scenario and for 2085.

The calculations in Table 7 are directly adapted from Ravindranath, Joshi, Sukumar, *et al.* (2006), where the authors run the BIOME4 model to compare the extent of area that is likely under each of the forest types under the present climate regime and future climate scenarios (see section "The vegetation model"). This output is originally reported in percentage terms, and we have converted it to approximate area figures in Mha, as shown in Table 7. Based on the grid size and data used in the BIOME4 model, it is difficult to calculate the exact areas for each of the forest types. Also, we cannot assume that a certain forest type will cover the total area of a grid, a part of which may be located inside. So as a proxy measure, we have divided the total forestland in India, 64 Mha

Table 7 Expected shifts in forest distribution

Current scenario		Area under B2 scenario (2085)										
Vegetation type	Current area (Mha)	TPXS	TPD/WL	WM	TPSD	TPS	TPEG	TMC	TMSW	CC	ET/M	CLDMX
TPXS	26.062	1.564	15.376	–	–	8.340	–	–	–	–	–	–
TPD/WL	17.281	–	9.331	–	0.173	1.382	6.048	–	–	–	–	–
WM	8.748	–	1.400	5.074	0.612	–	0.787	0.700	–	–	–	–
TPSD	5.135	–	0.051	–	0.359	–	4.673	–	–	–	–	–
TPS	2.851	–	0.827	–	–	1.882	0.114	–	–	–	–	–
TPEG	1.771	–	–	–	–	–	1.771	–	–	–	–	–
TMC	0.504	–	0.040	0.187	0.030	–	0.207	0.025	–	–	–	–
TMSW	0.475	–	0.005	0.408	–	–	–	–	0.052	–	–	–
CC	0.431	–	–	0.405	–	–	–	–	0.009	0.009	–	–
ET/M	0.407	–	–	0.098	–	–	–	0.004	–	0.106	0.077	0.057
CLDMX	0.337	–	–	0.266	–	–	–	–	0.020	–	–	–
Future area		1.564	27.031	6.437	1.175	11.604	13.600	0.729	0.081	0.115	0.077	0.057

Mha – million hectare; TPXS – tropical xerophytic shrubland; TPD/WL – tropical deciduous forest/woodland; WM – warm mixed forest/temperate evergreen/broadleaved; TPSD – tropical semi-deciduous forest; TPS – tropical savanna; TPEG – tropical evergreen forest; TMC – temperate conifer forest; TMSW – temperate sclerophyll woodland; CC – cool conifer forest; ET/M – evergreen taiga/montane forest; CLDMX – cold mixed forest

Source Ravindranath, Joshi, Sukumar, *et al.* (2006)

(Ravindranath 1998), proportionately on the basis of the number of grid points into the various forest types.

With an example we will illustrate how Table 7 can be read: WM forests are spread across India, covering 8.75 Mha of land. Roughly, 1.4 Mha of these forests will shift to TPD/WL, 0.61 Mha will shift to TPSD, 0.79 Mha will shift to TPEG, and 0.7 Mha will shift to TMC forest types. The remaining 5.1 Mha will remain as WM. Additionally, it can also be seen that the total spread of WM forests will decline from the current area of approximately 8.7–6.4 Mha. From Table 7, we can also identify which forest types are decreasing in area to a larger extent. These are TPXS, decreasing in area from the current 26.06 Mha to the projected 1.56 Mha. On the other hand, forests such as TPS are expected to increase in distribution, from the current 2.85 Mha to the projected 11.60 Mha in 2085.

Vulnerable areas

Once the magnitude of the impact is quantified, it becomes crucial to know whether such shifts in forest distribution are positive or negative. We thus need to set a criterion for selecting the vulnerable areas and continue our analysis focusing only on these. There might be changes in forest types, which can benefit biodiversity or livelihoods, but, on the other hand, some changes might be negative such as loss of NPP. We, therefore, have adopted a set of scientific criteria to identify the vulnerable areas: NPP, the biodiversity richness (BR) indicator, and vulnerability to forest fires (FF); ranks based on forest types are presented in Table 8. The criteria adopted are described below.

Net primary productivity

Forest biomes can be ranked in terms of NPP, depending on the annual growth of biomass. Ravindranath, Joshi, Sukumar, *et al.* (2006) provide in their study information about the NPP for the BIOME4 forests. With this information, we are able to rank forests in terms of NPP, and then identify positive and negative shifts accordingly.

Biological richness

BR, for biological richness or biodiversity richness, is an indicator of biodiversity that takes into consideration the ecosystem uniqueness, the diversity of species, the value of biodiversity, terrain complexity, and a disturbance index. Various vegetation types available in the

Table 8 Ranking of forest types based on the three criteria

Type of biome	Net primary productivity (NPP)	Biodiversity richness (BR)	Forest fire (FF)
Tropical xerophytic shrubland	10	11	1
Tropical deciduous forests/ woodlands	5	5	2
Warm mixed /temperate evergreen/ broadleaved	4	2	5
Tropical semi/moist deciduous	3	3	7
Tropical savannah	6	6	6
Tropical evergreen	1	1	11
Temperate conifer	2	8	3
Temperate sclerophyll woodland	9	10	8
Cool conifer	8	7	4
Evergreen taiga/montane	11	4	10
Cold mixed	7	9	9

Note 1 being maximum and 11 being minimum, the shifts to a higher number means a negative shift for NPP and BR, and positive for FF vulnerability.

terrain have been evaluated, and it is also dependent on the localized information. Finally, the BR is calculated as follows.

$$BR=\sum_{i=0}^{n} DI_i \times Wt_{i1} + TC_i \times Wt_{i2} + SR_i \times Wt_{i3} + BV_i \times Wt_{i4} + EU_i \times Wt_{i5} \quad (1)$$

where *DI* is the disturbance index; W_t is the weightage; *TC* is the terrain complexity; *SR* is the species richness; *BV* is the biodiversity value, and *EU* is the ecosystem uniqueness. Data on BR calculation and ranking was obtained from the National Remote Sensing Agency (NRSA 2007).

Forest fire vulnerability

The FF vulnerability indicator refers to the vulnerability to fire for a given forest type. This criterion has been adopted with the help of forest ecology experts by ranking forest types in terms of their vulnerability to FFs. The higher score (1) corresponds to the forest most vulnerable to fire (TPXS), while the lowest score (11) corresponds to the forest with low fire vulnerability. A shift from high vulnerability to lower vulnerability will, therefore, be categorized as a positive change.

Based on the three criteria shown in Table 8, we can now identify the shift direction, that is, whether the change from a particular forest type to another is positive or negative in terms of NPP, BR or FF vulnerability. Positive shifts will take place when increasing NPP, BR or decreasing FF vulnerability. In contrast, negative shifts will occur when decreasing NPP, BR or increasing FF vulnerability. Table 9 classifies each shift as positive or negative according to the forest rankings.

Table 9 Identification of the vulnerable areas

Forest type	Current area (Mha)	Shift to	Area in 2085 (Mha)	Impact direction		
				NPP	BR	FF
TPXS	26.062	TPD/WL	15.376	*	*	*
		TPS	8.340	*	*	*
TPD/WL	17.281	TPSD	0.173	*	*	*
		TPS	1.382	#	#	*
		TPEG	6.048	*	*	*
WM	8.748	TPD/WL	1.400	#	#	#
		TPSD	0.612	*	#	*
		TPEG	0.787	*	*	*
		TMC	0.700	*	#	#
TPSD	5.435	TPD/WL	0.051	#	#	#
		TPEG	4.673	*	*	*
TPS	2.851	TPD/WL	0.827	*	*	#
		TPEG	0.114	*	*	*
TPEG	1.771	TPEG	1.771			
TMC	0.504	TPD/WL	0.040	#	*	#
		WM	0.187	#	*	*
		TPSD	0.030	#	*	*
		TPEG	0.207	*	*	*
TMSW	0.475	TPD/WL	0.005	*	*	#
		WM	0.408	*	*	#
CC	0.431	WM	0.405	*	*	*
		TMSW	0.009	#	#	*
ET/M	0.407	WM	0.098	*	*	#

Continued...

Table 9 *Contd...*

Forest type	Current area (Mha)	Shift to	Area in 2085 (Mha)	Impact direction		
				NPP	BR	FF
		TMC	0.004	*	#	#
		CC	0.106	*	#	#
		CLDMX	0.057	*	#	#
CLDMX	0.337	WM	0.266	*	*	#
		TMSW	0.020	#	#	#

Mha - million hectare; TPXS - tropical xerophytic shrubland; TPD/WL - tropical deciduous forest/woodland; WM - warm mixed forest/temperate evergreen/broadleaved; TPSD - tropical semi-deciduous forest; TPS - tropical savanna; TPEG - tropical evergreen forest; TMC - temperate conifer forest; TMSW - temperate sclerophyll woodland; CC - cool conifer forest; ET/M - evergreen taiga/montane forest; CLDMX - cold mixed forest

Note * indicates positive impact direction; # indicates negative impact direction.

Sources Ravindranath, Joshi, Sukumar, *et al.* (2006); NRSA (2007)

As we can see from Table 9, TPD/WL forests will shift to TPSD, TPS, and TPEG. Currently, the forest type TPSD has a higher ranking in terms of NPP and BR, which means it has higher productivity and BR. For our purposes, we will call this a positive shift, as shown by the sign in the table. However, TPS is ranked lower than TPD/WL in terms of vulnerability to FF, and thus, the direction of this shift is considered as positive. Other entries could be read in a similar manner.

We consider vulnerable areas as those areas of the forest where the projected shift has been identified as negative by at least one of our three criteria. This leaves us with the vulnerable areas, as depicted in Table 10. Total area vulnerable for negative impacts due to climate change is computed as 6.2 Mha. Although this area only represents approximately 10% of the total current forest area in India, climate change impacts will be accompanied by high economic adaptation costs.

Table 10 Vulnerable areas

Current forest type	Forest type in 2085	Vulnerable area (Mha)	Impact direction		
			NPP	BR	FF
TPD/WL	TPS	1.382	*	*	#
WM	TPD/WL	1.400	*	*	*

Continued...

Table 10 *Contd...*

Current forest type	Forest type in 2085	Vulnerable area (Mha)	Impact direction		
			NPP	BR	FF
WM	TPSD	0.612	#	*	#
WM	TMC	0.700	#	*	*
TPSD	TPD/WL	0.051	*	*	*
TPS	TPD/WL	0.827	#	#	*
TMC	TPD/WL	0.040	*	#	*
TMC	WM	0.187	*	#	#
TMC	TPSD	0.030	*	#	#
TMSW	TPD/WL	0.005	#	#	*
TMSW	WM	0.408	#	#	*
CC	TMSW	0.009	*	*	#
ET/M	WM	0.098	#	#	*
ET/M	TMC	0.004	#	*	*
ET/M	CC	0.106	#	*	*
ET/M	CLDMX	0.057	#	*	*
CLDMX	WM	0.266	#	#	*
CLDMX	TMSW	0.020	*	*	*

Mha – million hectare; TPD/WL – tropical deciduous forest/woodland; WM – warm mixed forest/temperate evergreen/broadleaved; TPSD – tropical semi-deciduous forest; TPS – tropical savanna; TPEG –tropical evergreen forest; TMC – temperate conifer forest; TMSW – temperate sclerophyll woodland; CC – cool conifer forest; ET/M – evergreen taiga/montane forest; CLDMX – cold mixed forest; NPP – net primary productivity; BR – biodiversity richness; FF – forest fire

Note * Indicates positive impact direction; # indicates negative impact direction.

Sources Ravindranath, Joshi, Sukumar, *et al.* (2006); NRSA (2007)

Identification of adaptation options

From a review of the literature, a set of potential adaptation options for Indian forests is presented in Table 4. From these, we select the adaptation options related to shifts in forest distribution, which correspond to changes in productivity, biodiversity, and FF. The result is the following set of adaptation options (Table 11).

Table 11 Adaptation options for forest biomes in India

Adaptation option	Specific adaptation measure
A1	Fauna and flora migration/biological corridors
A2	Adaptation of the protected areas system
A3	Plantation of drought-tolerant species to avoid fragmentation
A4	Allow natural adaptation: species dispersal (management)
A5	Enable population mortality and colonization
A6	Negotiation of conservation easements as protection from development
A7	Community instruments to reduce forest dependency
A8	Flexibility in silvicultural management
A9	Modification of rotation times
A10	Plantation of productive species
A11	Forest fire prevention programmes
A12	Rehabilitation of degraded forest

Sources FAO (2008); GoI (2008); Heller and Zavaleta (2009).

For these measures, we now need to apply four selection criteria—relevance, effectiveness, scale of action, and feasibility.

Relevance

The relevance of adaptation options is based on the direction of the impact. The information gathered in the previous steps will be subsequently used to match the adaptation options pertinent to productivity losses, biodiversity losses, and FF risks. For a particular forest type, if all or a combination of these factors is applicable, then the relevant set of options will be applied. The selection of the relevant adaptation options has been made case-specific, going through each vulnerable area and using expert opinion to recommend the adaptation options. The criteria that have been followed to choose a given adaptation option are summarized in Box 1.

Furthermore, we have divided the choice of adaptation options into two baskets: Scenario 1 for the lower bound (minimum action with most urgent options) and Scenario 2 for the upper bound (with all desirable options). For the worst case scenario or Scenario 1, marked as # in Table 12, the most urgent measures to address the impacts are chosen. For the best case scenario or Scenario 2, marked as # and * in the table, all the possible adaptation options are chosen, including the previous ones. The selection of relevant adaptation options has been done based on expert opinion and is presented as Table 12.

As an example, the area under TPD/WL, which is expected to shift towards TPS, will require adaptation options A3, A7, A10, and A11 as the most urgent options (adaptation Scenario 1). Among all desirable options, adaptation Scenario 2 for TPD/WL will be implementing A1 and A3–A11 options for the entire stretch of 1.38 Mha of this forest type that has been identified as vulnerable.

Box 1

Criteria for selecting the relevant adaptation options

A1 - If the affected forests have important fauna species.

A2 - If vulnerability of the forest increases due to external factors (invasive species, human intervention/pressure). This has been followed in general terms since information per region is not available in this approach.

A3 - If shifts are towards a drier forest type.

A4 - If shifts are to a more desirable forest in terms of productivity and biodiversity.

A5 - If shift is happening towards a more desirable forest in terms of biodiversity.

A6 - If shifts are in the direction of a higher vulnerability to human interventions and human pressure.

A7 - When a shift is towards forests that produce a lower quantity of anthropogenic products or towards highly human-dependent forest change.

A8 and A9[a]

A10 - If shifts are towards drier forest types with impacts in their productivity.

A11 - If shifts are towards drier forest types and vulnerability are expected to increase due to human pressure.

A12 - If vulnerability of the forest increases owing to external factors (invasive species, human intervention/pressure).

[a] Criteria could not be established in these cases since information was not available for public forests on a national scale.

Table 12 Relevance of the adaptation options

Current forest type	Forest type in 2085	Vulnerable area (Mha)	Adaptation options											
			A1	A2	A3	A4	A5	A6	A7	A8	A9	A10	A11	A12
TPD/WL	TPS	1.382	*	†	#	*	*	*	#	*	*	#	#	†
WM	TPD/WL	1.400	*	*	#	*	†	*	#	*	*	#	#	*
WM	TPSD	0.612	*	*	#	*	*	†	†	†	†	†	#	*
WM	TMC	0.700	†	†	#	*	*	†	#	†	†	†	#	†
TPDS	TPD/WL	0.051	*	†	#	*	*	†	#	*	*	#	#	†
TPS	TPD/WL	0.827	†	†	†	†	†	†	†	†	†	†	†	†
TMC	TPD/WL	0.040	†	†	#	#	#	*	*	*	*	#	#	†
TMC	WM	0.187	†	†	†	#	#	*	*	*	*	#	†	†
TMC	TPSD	0.030	†	†	†	#	#	*	*	*	*	*	†	†
TMSW	TPD/WL	0.005	†	†	†	†	†	†	†	†	†	†	†	†
TMSW	WM	0.408	†	†	†	†	†	†	†	†	†	†	†	†
CC	TMSW	0.009	*	*	*	*	*	*	*	*	*	*	*	*
ET/M	WM	0.098	†	†	†	†	†	†	†	†	†	†	†	†
ET/M	TMC	0.004	†	*	*	†	*	†	#	†	†	†	*	*
ET/M	CC	0.106	*	*	#	†	*	†	†	†	†	†	#	*
ET/M	CLDMX	0.057	*	*	#	†	*	†	†	†	†	†	#	*
CLDMX	WM	0.266	†	†	†	†	†	†	†	†	†	†	†	†
CLDMX	TMSW	0.020	*	*	#	*	*	*	*	*	*	*	#	*

Mha – million hectare; TPXS – tropical xerophytic shrubland; TPD/WL – tropical deciduous forest/woodland; WM – warm mixed forest/temperate evergreen/broadleaved; TPSD – tropical semi-deciduous forest; TPS – tropical savanna; TPEG –tropical evergreen forests; TMC – temperate conifer forest; TMSW – temperate sclerophyll woodland; CC – cool conifer forest; ET/M – evergreen taiga/montane forest; CLDMX – cold mixed forest

# Indicates options for the worst case scenario or Scenario 1; * indicates additional adaptation options for the best case scenario or Scenario 2; † indicates that adaptation options are not suitable for the expected impacts.

Source Ravindranath, Joshi, Sukumar, *et al.* (2006) and own elaboration.

Effectiveness

The effectiveness of the adaptation options depends on the parameter needed to determine the success of the adaptation options in reducing climate impacts. Unfortunately, due to the lack of information on the effectiveness of the adaptation measures individually, we cannot quantify what per cent of the impact will be offset in our case study. Therefore, we do not have information on the success of the adaptation options and will assume that each relevant and feasible option has to be implemented in the total vulnerable area.

Scale

The scale for the adaptation options will be the area in which the adaptation policies should be implemented. As such, we consider that each adaptation option fulfilling these four criteria will be implemented in the vulnerable areas as identified earlier. These areas are already calculated at this point and can be taken from the column giving the areas of the shift in Table 12.

Feasibility

The feasibility of the adaptation options depends on the existing policies and ongoing programmes similar to these actions in the field. This is mainly due to the availability of cost estimates for the actions, which ideally should be based on similar existing or projected programmes. We will come back to this point later in this chapter, describing the information on unit costs. Based on the available information, only the feasible adaptation options will be included in the adaptation cost estimates.

Identification of unitary costs for the adaptation measures

In order to obtain the unitary costs for the adaptation measures identified, we do a profound review on the Indian national programmes on forest management. With this, we identify those measures, the objectives of which match the list of potential adaptation measures that we have identified. From these sources, we select the costs given in the Eleventh Five-year Plan presented by the Working Group on Forests (GoI 2007) and the National Forestry Action Programme (NFAP), as they provide information about the area, time frame, and typologies of costs, fitting our adaptation options. From the available programmes in the Eleventh Plan, and with the information available on the forest coverage (in hectares) and on the time frame of the programmes, we calculate the annual cost per hectare for each measure and link it to the potential adaptation options. Costs have been divided into capital and recurrent, and have been converted to 2009 US dollars. Rupees have been updated to 2009 value, with the consumer price index from the World Economic Outlook database from the International Monetary Fund (IMF) (WEO 2009), and converted to US dollars with the most recent exchange rate of September 2009 and also based on the exchange rate of the IMF (2009). The net present values have been calculated for a 10-year period at an interest rate of 12%. Table 13 shows the feasible adaptation options for which detailed costs have been identified.[1]

In order to estimate the costs of these adaptation actions, we have constructed an implementation timeline for each adaptation option. As the impacts are expected in the long term (year 2085), these sets of adaptation options will be implemented in the future. Also, because short-term climatic impacts on forests are mainly related to increases in NPP, we expect these negative impacts to become relevant in the mid-term. Therefore, we will estimate the adaptation costs by applying these options to avoid the expected impacts in 2085. We assume that the different adaptation options will be implemented in the period 2075–85.

Table 14 summarizes the adaptation costs estimated for this period. The interval corresponds to the adaptation scenarios (Scenario 1 and Scenario 2, correspondingly) and costs are in 2009 US dollars on an annual basis. The final column resumes the expected cost for the 10-year period. For more detailed information on the cost calculations, see Annexure II.

For the entire period, we can add up the annual costs and obtain total adaptation costs, which amount to $1781.71 million in

---

[1] Cost figures for the rest of adaptation options were not feasible due to the lack of information on the magnitude and distribution of migration corridors (A1), no information on species distributions to design protected areas increments (A2), similarity between A4 and A5 (we include A4 since information on costs is available), lack of information on conservation easements in private lands (A6), and details on the silviculture managements, which are also dependent on the region and property regime (A8 and A9).

Table 13 Costs of the adaptation options (prices in 2009, $/ha)

Adaptation option		Capital (NPV)	Year 1	Year 2	Year 3	Source
A3	Plantation of drought-tolerant species	86.10	90.81	25.95	25.95	GoI (2007)
A4	Allowing natural adaptation	3.10	17.53	17.53	17.53	GoI (2002)
A7	Reducing forest dependency	23.48	16.95	16.95	16.95	Government of Haryana (2003)
A10	Plantation of productive species	86.10	90.81	25.95	–	GoI (2007)
A11	Forest fire prevention	–	2.63	2.63	2.63	Karnat (2008)
A12	Rehabilitation of degraded forest	167.96	178.39	89.52	89.52	Government of Haryana (2003)

NPV – net present value

Sources Government of Haryana (2003); GoI (2002, 2007); Karnat (2008)

adaptation Scenario 1 and $3050.76 million in adaptation Scenario 2. These costs are divided into capital and recurrent costs in Table 15.

Adaptation costs in the shorter term: 2030 estimates

In the present chapter, we have developed and applied a detailed methodology for estimating the cost of adaptation for Indian forests. As climatic impacts on forest could only be quantified for 2085, our cost estimates refer to the expected impacts for 2085.

For the purpose of the present study, we also want to present an estimate of the adaptation costs for India in the shorter term, in line with the other sectors. To do this, as 2030 impacts are unknown, and given that forests are expected to increase in productivity in the short term due to the fertilization effect (DEFRA 2005), we will base our calculations on non-regret adaptation options. Based on the judgments of forest experts in The Energy and Resources Institute (TERI), we have reviewed the Tenth Five-year Plan and the current (Eleventh) Five-year Plan presented by the Working Group on Forests (GoI 2007) to identify those policies that are more suitable

Table 14 Estimated annual adaptation costs for the period 2075–85 (in million $)

Cost categories	2076	2077	2078	2079	2080	2081	2082	2083	2084	2085
Capital	78.16–118.30	64.04–103.04	78.16–118.30	64.04–103.04	78.16–118.30	64.04–103.04	78.16–118.30	64.04–103.04	78.16–118.30	64.04–103.04
Management	1.15–1.15	79.25–127.44	98.97–174.43	128.89–232.44	118.70–221.43	128.89–232.44	118.70–221.43	128.89–232.44	138.42–268.42	128.89–232.44
Total	79.31–119.45	143.29–230.47	177.13–292.73	192.92–335.48	196.85–339.73	192.92–335.48	196.85–339.73	192.92–335.48	216.58–386.72	192.92–335.48

Note The lower and upper bound estimates correspond to the adaptation Scenario 1 and Scenario 2.

Table 15 Total adaptation costs for the period 2075–85 (in 2009 million $)

Cost categories	Adaptation costs
Capital	710.98–1106.71
Management	1070.72–1944.06
Total	1781.71–3050.76

Note The lower and upper bound estimates correspond to the adaptation Scenario 1 and Scenario 2.

for climate adaptation. These non-regret actions are considered as the adaptation level to be achieved for 2030. Table 16 relates to the costs of the ongoing schemes, as well as the new schemes to be implemented by the Eleventh Five-year Plan. In the last column, the annual costs for the climate-adaptation-related schemes that we are considering are shown.

Table 16 Costs of the proposed schemes for forest development under the Tenth and Eleventh plan outlay, and selected non-regret options (2009 million $)

Ongoing schemes	Tenth Five-year Plan outlay	Eleventh Five-year Plan outlay	Non-regret options (million $/year)
NAP scheme[a]	284.71	1890.19	70.88
EDF scheme	16.54	23.63	–
NAEB scheme	29.53	47.25	–
Gregarious Flowering of Bamboo in the North East	10.11	35.44	–
IFPS scheme[a]	106.32	165.39	33.08
Conservation of Mangroves, Coral Reefs and Wetlands scheme	12.76	16.54	–
Biodiversity and Conservation scheme	2.84	7.09	–
Scheme for capacity building in the forest sector (IGNFA, DFE, IFS officers)	12.05	40.40	–
Scheme for research and training support	64.98	292.51	–
New schemes			
Gram Van Yojana scheme	–	2268.23	–
Scheme for forest extension and marketing supports	–	236.27	–
Scheme for augmentation of bamboo resources	–	236.27	–
Conservation scheme of medicinal plant resources	–	482.00	–

Continued...

Table 16 *Contd...*

Ongoing schemes	Tenth Five-year Plan outlay	Eleventh Five-year Plan outlay	Non-regret options (million $/year)
Special forest protection scheme for Naxalite areas	–	118.14	–
Forest Information Management and Resource Assessment Scheme[a]	–	70.88	11.81
Forest Infrastructure Development Scheme[a]	–	141.76	28.35

NAP - National Afforestation Programme; EDF - Eco-Development Forces; NAEB - National Afforestation and Eco-Development Board; IFPS - Integrated Forest Protection; IGNFA - Indira Gandhi National Forest Academy; DFE - Directorate of Forest Education, IFS - Indian Forest Service.
[a] Non-regret options selected for adapting to climate change in 2030.

Based on Table 16, annual costs for the non-regret adaptation options to be implemented from 2010 to 2030 amount to $144.13 million per year. This would be the cost of adapting forests to climate-change-based non-regret options because of the lack of information on forest-related climatic impacts in the shorter term.

Summary of the results

Table 17 summarizes the adaptation costs that have been calculated for the Indian forests. First column corresponds to the short-term adaptation costs based on climate-oriented non-regret options, obtained from current forest five-year plans (GoI 2007). The next three columns correspond to the adaptation costs following the detailed methodology developed in this chapter, based on the projected impacts (new programme). A new programme is designed to be implemented with the adaptation options satisfying the criteria. As a result, we can see as to how costs will be important in the long term, depending on the adaptation scenario chosen, varying from $192.92 million per year to $335.48 million per year in 2085.

Table 17 Total adaptation costs in Indian forests in the short and in the long run (annual costs in 2009 million $)

Annual expenditure	Ongoing programmes	New programme		
	2010-2030	2076	2080	2085
Capital	144.13	78.16-118.30	78.16-118.30	64.04-103.04
Recurrent		1.15-1.15	118.70-221.43	128.89-232.44
Total costs net present value	144.13	79.31-119.45	196.85-339.73	192.92-335.48

Conclusion

In this chapter, a new methodology for estimating adaptation costs for ecosystems has been presented and applied to forest ecosystems in India. Recognizing the urgent need for information on the costs of adaptation and due to the general and very few estimates of adaptation costs for ecosystems existing in the literature, the aim of the chapter has been to develop a new methodological framework for estimating the costs of adaptation for ecosystems. This methodology contributes significantly to the existing literature, mainly by including a set of adaptation options, by identifying the areas that are vulnerable to climate change, and by considering the magnitude of the impacts and the adaptation level that is required to avoid this impact.

The methodological approach presented in this chapter has been designed to be applied to different types of ecosystem and is thus intended to be versatile. Forest ecosystems have been the focus of the analysis that illustrates the methodology adopted, wherein adaptation costs have been estimated for India. As a result, adaptation costs for forests in India are in the range of $192.92 million and $335.48 million per year in 2085, depending on the adaptation scenario. These estimates are much lower than the adaptation costs reported in the literature such as the one illustrated by Blamford, Bruner, Cooper, *et al.* (2002) or Berry (2007), but these represent a global dimension, and the extent of adaptation needed is not measured or projected. In the case of the present analysis, estimates are at the country level, including different adaptation measures and employing scientific evidence on the expected impacts of climate change on Indian forests. This strength of the study is also a limitation because our methodology relies on the available scientific data on climatic change impacts, and this information is for 2085 under the B2 scenario.

As adaptation strategies are also needed in the shorter term, we have estimated adaptation costs for 2030 based on non-regret options given in current forest plans. We lack information on what the impacts on Indian forests will be in the short term. Moreover, productivity of forests is expected to increase during the next decades as an effect of climate change. To calculate the adaptation costs for this shorter period, we have selected climate-oriented actions being taken or projected in India. Based on these, the estimated annual cost of forest adaptation for 2030 amounts to $144.13 million.

This study has, however, a scope of possible improvements, which we would like to highlight for future applications. It would have been desirable to take a state-wise approach for India so that impacts and adaptation options could be planned and costs estimated state by state. Including more climate IPCC scenarios is also necessary, although the availability of data is crucial for this. Future work will demonstrate the potential and large-scale application of this methodology for estimating adaptation costs in other types of ecosystems such as wetlands or coastal, at different scales and in different countries around the world.

ANNEXURES

Annexure 1

Distribution of the forest biomes

In Table I, the corresponding regions in India for the vegetation types are shown. The first column corresponds to the BIOME4 classification. The second column shows the matching classification using multi-temporal IRS wide field sensor (WiFS) data,[2] while the third column includes the regions where such vegetation types exist in India. This mapping is done based on comparisons between various studies, Champion and Seth Classification, and expert[3] ecologist opinion.[4]

Table I Types of forest biomes

Forest type	Vegetation type	Regions in India
TPXS	Shrubs (Sc)	Central Highlands and areas of South India
TPD/WL	TDDF	Rajasthan, Madhya Pradesh, Maharashtra, Orissa, Uttar Pradesh, Karnataka, Andhra Pradesh, Tamil Nadu, Punjab, Haryana
WM/temperate evergreen/ broadleaved	–	North East India, West Bengal, Uttarakhand (1500–3000 m)
TPSD	TMDF	Andaman and Nicobar Islands, Himalayan Foothills, East of Western Ghats, Chota Nagpur, North West Indian hills
TPS	Grasslands	Gujarat, Central India
TPEG	TEF	Western Ghats, Upper Assam, Arunachal Pradesh, Andaman and Nicobar, Tamil Nadu, Andhra Pradesh, Himachal Pradesh
TMC	TCF	Eastern and Western Himalayas (1800–3000 m)
TMSW	TMF	Himachal Pradesh, Jammu and Kashmir, Higher Uttarakhand
CC	CF	Central and Western Himalayas, NE hills

Continued...

[2] Joshi, Roy, Singh, *et al.* (2006)
[3] P K Joshi, National Remote Sensing Agency (NRSA) and TERI University
[4] Champion and Seth (1968)

Table I *Contd...*

Vegetation types	Vegetation types	Regions in India
ET/M		Nilgiri, Anamalai and Tirunelveli hills above 1500 m.
CLDMX	TMF	Himachal Pradesh, Jammu and Kashmir, Higher Uttaranchal

TPXS - tropical xerophytic shrubland; TPD/WL - tropical deciduous forest/woodland; TDDF - tropical dry deciduous forest; WM - warm mixed forest/temperate evergreen/broadleaved; TPSD - tropical semi-deciduous forest; TMDF - temperate deciduous forest; TPS - tropical savanna; TPEG - tropical evergreen forest; TMC - temperate conifer forest; TMSW - temperate sclerophyll woodland; CC - cool conifer forest; TEF - tropical evergreen forest; TCF - temperate conifer forest; TMF - temperate mixed forest; CF - conifer forest; ET/M - evergreen taiga/montane forest; CLDMX - cold mixed forest

Annexure II

Calculation of adaptation costs for 2075–85

Adaptation options to avoid shifts in forest distribution expected in India in 2085 have been designed to be implemented over a 10-year period previous to the year at which impacts are expected. The time line of the adaptation options employed based on the calculation of costs follows Table I. As an illustration, adaptation option A3 will be implemented in a 10-year period in an area to be determined. The implementation will include a capital investment and three years of management. The total area is divided by 10, so each year a parcel is planted with the incurred adaptation costs (in green). Then, in the second year another parcel is planted (in green), and the first plantation now needs management costs (in orange), so in the next year we will have plantation costs on 1/10 of the area and management costs also in 1/10 of the affected area.

Table I Time line for the adaptation options

	Year	2076	2077	2078	2079	2080	2081	2082	2083	2084	2085
Adaptation option	Area portion										
A3	1/10										
	1/10										
	1/10										
	1/10										
	1/10										
	1/10										
	1/10										
	1/10										
	1/10										
	1/10										
	1/10										
	1/10										
	1/10										

Continued...

Table I *Contd...*

	Year	2076	2077	2078	2079	2080	2081	2082	2083	2084	2085
Adaptation option	Area portion										
A11	1										
	1/10										
	1/10										
	1/10										
	1/10										
A12	1/10										
	1/10										
	1/10										
	1/10										
	1/10										
	1/10										
	1/6										
	1/6										
	1/6										
A7	1/6										
	1/6										
	1/6										
	1/10										
	1/10										
	1/10										
	1/10										
	1/10										
A10	1/10										
	1/10										
	1/10										
	1/10										
	1/10										
A11	1										
	1/10										
	1/10										
	1/10										

Continued...

Table I *Contd...*

	Year	2076	2077	2078	2079	2080	2081	2082	2083	2084	2085
Adaptation option	Area portion										
	1/10										
	1/10										
	1/10										
	1/10										
	1/10										
	1/10										
	1/10										

Following this implementation pattern, the adaptation costs for each shift in forest area are presented in Table II.

Table II Total adaptation costs for each vulnerable area and adaptation scenario (2009 in million$)

Current forest type	Future forest type	Shift	Capital costs		Recurrent costs		Total costs	
		Mha	S1	S2	S1	S2	S1	S2
TPD/WL	TPS		265.58	269.87	396.06	456.64	661.64	726.51
WM	TPD/ WL		269.04	508.53	401.21	887.87	670.25	1396.40
WM	TPSD		52.70	157.39	77.04	289.78	129.73	447.17
WM	TMC		74.25	76.42	114.33	145.02	188.58	221.44
TPSD	TPD/ WL		9.80	9.96	14.62	16.85	24.42	26.81
TMC	TPD/ WL		0.00	0.00	0.00	0.00	0.00	0.00
TMC	WM		7.01	7.81	11.72	13.22	18.73	21.03
TMC	TPSD		16.68	20.41	31.25	38.25	47.93	58.66
TMSW	TPD/ WL		0.09	3.28	1.32	6.14	1.41	9.41
TMSW	WM							
CC	TMSW							
ET/M	TMC			3.27		5.71		8.98
ET/M	CC							

Continued...

Table II *Contd...*

Current forest type	Future forest type	Shift	Capital costs		Recurrent costs		Total costs	
		Mha	S1	S2	S1	S2	S1	S2
ET/M	CLDMX		0.08	1.10	0.15	1.87	0.23	2.96
CLDMX	WM		9.13	26.93	13.34	45.54	22.47	72.47
CLDMX	TMSW		4.91	14.48	7.18	24.49	12.08	38.97
Total costs			709.26	1099.44	1068.20	1931.37	1777.47	3030.81

S1 - adaptation Scenario 1 or worst case scenario; S2 - adaptation Scenario 2 or best case scenario; Mha - million hectare; TPD/WL - tropical deciduous forest/woodland; TPS - tropical savanna; WM - warm mixed forest/temperate evergreen/broadleaved; TPSD - tropical semi-deciduous forest; TMC - temperate conifer forest; TMSW - temperate sclerophyll woodland; CC - cool conifer forest; ET/M - evergreen taiga/montane forest; ; CLDMX - cold mixed forest

References

Berry P. 2007. ***Adaptation Options on Natural Ecosystems. A Report to the UNFCCC Secretariat Financial and Technical Support Division***. Oxford, UK: Environmental Change Institute

Balmford A, Bruner A, Cooper P, Costanza R, Farber S, Greem R E, Jenkins M, Jefferies P, Jessamy V, Madden J, Munro K, Myers N, Naeem S, Paavola, Rayment M, Rosendo S, Roughgarden J, Trumper K, and Turner R. T. 2002. **Economic reasons for conserving wild nature**. *Science* **297:** 950–953.

Champion H and Seth S. 1968. ***A Revised Survey of Forest Types of India***. New Delhi: New Delhi Government Publication.

Cruz R, Harasawa H, Lal M, Wu S, Anokhin Y, Punsalmaa, B, Honda Y, Jafari M, Li C. 2007. **Asia.** In *Climate Change 2007: impacts, adaptation, and vulnerability*, pp. 469–506, edited by M Parry, O Canziani, J Palutikof, P van der Linden, and C Hanson. Cambridge, UK: Cambridge University Press

DEFRA (Department for Environment, Food and Rural Affairs). 2005. ***Climatic Change Impacts on Forestry in India***. Keysheet 7. Indian Institute of Science (IISc) Bangalore, India

Easterling W, Aggarwal P, Batima P, Brander K, Erda L, Howden A, Kirilenko J, Morton J-F, Soussana S, Schmidhuber, Tubiello F. 2007. **Food, fibre, and forest Products.** In *Climate Change 2007: impacts, adaptation, and vulnerability*, pp. 273–313, edited by M Parry, O Canziani, J Palutikof, P van der Linden, and C E Hanson. Cambridge, UK: Cambridge University Press

EEA (European Environmental Agency). 2008. **The economics of ecosystems and biodiversity**. Details available at <http://ec.europa.eu/environment/nature/biodiversity/economics>

FAO (Food and Agriculture Organization). 2008. **Climate change and biodiversity for food and agriculture**. [Technical Background Document from the Expert Consultation]

Ferraro P and Kiss A. 2002. **Direct payments to conserve biodiversity.** *Science* **298**: 1718–1719.

Fischlin A, Midgley G, Price J, Leemans R, Gopal B, Turley C, Rounsevell M D A, Dube O P, Tarazona J, Velichko A A. 2007. **Ecosystems, their properties, goods, and services**. In *Climate Change 2007: impacts, adaptation, and vulnerability*, pp. 211–272, edited by M Parry, O Canziani, J Palutikof, J van der Linden, and C Hanson, Cambridge, UK: Cambridge University Press

GoI (Government of India), 2002. **10th Five-year Plan (2002–07)**. Details available at <http://planningcommission.gov.in/plans/planrel/fiveyr/index10.html>

GoI (Government of India), 2007. ***Report of Working Group on Forests for the Eleventh Five-year Plan (2007–12)***. New Delhi: Planning Commission, GoI

GoI (Government of India), 2008. ***India National Action Plan for Climate Change***. Ministry of Environment and Forests, Government of India. Details available at <http://pmindia.nic.in/Pg01-52.pdf>

Government of Haryana. 2003. ***Integrated Natural Resource Development For Economic Growth in Rural Haryana Through Forestry Programmes***. Haryana: Forest Department, Government of Haryana

Heller N E and Zavaleta E S. 2009. **Biodiversity management in the face of climate change: a review of 22 years of recommendations.** *Biological Conservation* **142**: 14–32

IMF (International Monetary Fund). 2009. **Representative rates for selected currencies.** Details available at <www.imf.org/external/np/fin/data/param_rms_mth.aspx>

IPCC (Intergovernmental Panel on Climate Change). 2007. **Technical summary**. In *Climate Change 2007: Impacts, Adaptation, and Vulnerability*, pp. 23–78, edited by M Parry, O F Canziani, J P Palutikof, P J van der Linden, C E Hanson. Cambridge, UK: Cambridge University Press.

James A, Gaston K J, and Balmford A. 2001. **Can we afford to conserve biodiversity?** *BioScience* **51**(1): 43–52

Joshi P K, Roy P, Singh S, Agrawal S, Yadava D. 2006. **Vegetation cover mapping in India using multi-temporal IRS Wide Field Sensor (WiFS) data.** *Remote Sensing of Environment* **103**: 190–202

Karnat M. 2008. **Forest fire management under the modern forest fire control project in Chandrapur District, Maharashtra**. In *Rethinking Forest Fires*, edited by R Pai, A J Hiremath and U Kant (eds), [Proceedings of the National Workshop on Forest Fires, 13–14 November 2007, New Delhi. Ministry of Environment and Forests, Government of India. Ashoka Trust for Research in Ecology and the Environment (ATREE)]. 153 pp.

MEA (Millennium Ecosystem Assessment). 2005. ***Ecosystems and Human Well-Being: biodiversity synthesis.*** Washington, DC: World Resources Institute. 36 pp.

Naidoo R and Ricketts T. 2006. **Mapping the economic costs and benefits of conservation**. *Public Library of Science Biology* **4**(11): 9495-9500.

Negi S S. 1993. ***Biodiversity and its Conservation in India***. New Delhi: Indus Publishing Company. 243 pp.

NRSA (National Remote Sensing Agency). 2007. ***Biodiversity Characterisation at Landscape Level in Central India and West Bengal Using Satellite Remote Sensing and Geographical Information System***. Dehradun: Indian Institute of Remote Sensing, NRSA, Department of Space, Government of India. 542 pp.

Pagiola S, von Ritter K, and Bishop J. 2004. ***Assessing the Economic Value of Ecosystem Conservation***. Washington, DC: World Bank. [Environment Department Paper No. 101]

Parry M, Arnell N, Berry P, Dodman D, Fankhauser S, Hope C, Kovats S, Nicholls R, Satterthwaite D, Tiffin R, Wheeler T. 2009. ***Assessing the Costs of Adaptation to Climate Change: a review of the UNFCCC and other recent estimates.*** London: International Institute for Environment and Development and Grantham Institute for Climate Change.

Paterson J S, Araujo M B, Berry P M, Piper J, Rounsevell M. 2008. **Mitigation, adaptation, and the threat to biodiversity.** *Conservation Biology* **22**(5): 1352–1355

Pimm S L and Raven P. 2000. **Extinction by numbers.** *Nature* **403**: 843–845.

Ravindranath N H. 1998. **Climate change and tropical forests in India**. *Climatic Change* **39**: 563–581

Ravindranath N H, Joshi N, Sukumar R, Saxena A. 2006. **Impact of climate change on forest in India**. *Current Science* **90**(3): 354–361

Ravindranath N H, Sukumar R, and Deshingkar P. 1997. ***Climate change and forests: impacts and adaptation***. In *Regional Assessment for the Western Ghats, India*. Stockholm: Stockholm Environment Institute

Ravindranath N and Murthy I K. 2003. **Clean Development Mechanism and forestry projects: strategy for operationalization in India.** *The Indian Forester* **129**(6): 691–706

Singh J, Singh S, and Gupta S. 2006. ***Ecology, Environment and Resource Conservation***. New Delhi: Anamaya Publishers. 688 pp.

Stern N S. 2006. ***Stern Review: the economics of climate change***. London: HM Treasury. 713 pp.

Thomas C D H. 2004. **Extinction risk from climate change**. Letters to Nature, *Nature* **427**: 145–148

UNDP (United Nations Development Programme). 2007. ***Human Development Report 2007/08***. New York: Palgrave Macmillan. 399 pp.

UNFCCC (United Nations Framework Convention on Climate Change). 2007a. ***Climate Change: impacts, vulnerabilities, and adaptation in developing countries***. Bonn, Germany: UNFCCC. 68 pp.

UNFCCC (United Nations Framework Convention on Climate Change). 2007b. ***Investments and Financial Flows to address Climate Change.*** Bonn, Germany: Climate Change Secretariat. Details available at <unfccc.int/cooperation_and_support/financial_mechanism/items/4053.php>

WEO (World Economic Outlook). 2009. **World Economic Outlook Database**. Details available at <www.imf.org/external/pubs/ft/weo/2009/01/weodata/index.aspx>

Zaunberger K, Stefan A, and Ladislav M. 2008. **The climate change: biodiversity nexus key to co-benefit approaches**. Details available at <http://circa.europa.eu/Public/irc/env>. [Draft paper for the European Commission, Directorate General for Environment.]

four

Agriculture

Suruchi Bhadwal,[a] *Sambita Ghosh,*[b] *and Julia Martin-Ortega*[c]

Background

India's agriculture sector contributes significantly to the employment of people, with 70% of the population dependent on it either directly or indirectly. In addition, it caters to the needs of a number of agro-processing industries and contributes significantly to the exports from the country (GoI 2008).

An important limiting factor to be borne in mind is that two-thirds of the land area in India is classified as arid or semi-arid, and a major part of the agriculture (68% of the net cultivated area) in the country is rain-fed (GoI 2001). The distribution of rainfall and potential evapotranspiration rates determine the types of crops grown across the country. At present, a large percentage of the cultivated area is under foodgrain production.

Post Independence, the country has transformed itself from a food-deficit country to a food-secure country. Production of foodgrains has increased four times, from roughly 50 million tonnes (MT) in 1950/51 to 217.3 MT in 2006/07 (Table 1). However, the current yield of foodgrains is still short of the potential yields that can be achieved, and there is tremendous scope to advance it further. For example, the production of wheat and rice has been observed to increase from 2 tonnes/hectare (t/ha) in 1961 to 5 t/ha in 2000,

[a] Fellow, Climate Change, Centre for Global Environment Research, The Energy and Resources Institute (TERI), New Delhi, India.

[b] Research Associate, Climate Change, Centre for Global Environment Research, The Energy and Resources Institute (TERI), New Delhi, India.

[c] Basque Centre for Climate Change (BC3), Bilbao, Spain.

and the aggregate rice production has increased from 20 MT to 90 MT (IPCC 2007). Large-scale regional variations exist; for example, the productivity of wheat in Maharashtra is reported to be 1.3 t/ha, while that in Punjab is 4.1 t/ha.

The overall increase in yields is attributed to the Green Revolution, which, through the introduction of hybrid varieties and emphasis on increased inputs of fertilizers, pesticides, and water, has contributed to increase in agricultural production. In fact, much of the success in agricultural production and observed productivity enhancements has been based on agricultural intensification, with the net cultivated area remaining stable over the last three decades or so. At present, India contributes to a major share of the world's agricultural production, ranking third globally in cereal production (MoA 2004).

Though this progress in production rates is commendable, there has been a decline in the growth rate of some crops in certain regions. This is largely attributed to nutrient depletion and environmental degradation. Productivity of some of the cropping systems has either stagnated or is showing a decline in the trends. While growth rates of yields have become static in the case of rice, those for wheat have gone down in recent times. While, on the one hand, there has been progress, on the other, there are concerns related to this intensification process and its sustainability in the long run. These include degradation of land resources, contamination of groundwaters and surface waters, decline in crop diversity, and increased waterlogging and soil salinity. There are also other threats related to changes in land use and land cover with the expansion of urbanization. The number of peri-urban areas has seen a drastic increase over the last decade or so. Also, challenges due to the conversion of agricultural land for biofuel plantations are emerging.

Literature survey on impacts of climate change on agriculture and costs of adaptation

In addition to the above-mentioned issues, climate change is likely to exacerbate the negative trends further. Under an enhanced greenhouse condition, the variability of monsoons is projected to increase. This will result in recurring droughts or floods, which would affect agricultural production and increase the vulnerability of the agrarian population. Acute water shortage conditions, combined with thermal stress, would adversely affect the yield of many food crops in the region.

Impacts on agriculture

Climate change is likely to affect agriculture through direct and indirect effects on crops, livestock, the microbial population, soils, and pests.

While direct positive impacts are associated with the carbon fertilization effect, indirect effects are associated with the losses due to increase in frequency and intensity of extreme events, changes in the water availability situation and soil moisture status, increase in pest and disease incidence, and decline in arable land. Furthermore, increase in temperatures can result in the reduction in crop duration, increase in crop respiration, alterations in metabolic processes, changes in survival and distribution of pests and diseases, alterations in the decomposition of organic matter, changes in nutrient uptake and removal by soils, decrease in fertilizer use efficiency, and increase in evapotranspiration.

Specific regions and certain groups of farmers in the country, particularly those farming on marginal, rain-fed lands, are likely to suffer significant damages as a result of climate change (Dinar, Mendelsohn, Evenson, *et al.* 1998). West Coast agricultural regions, including Gujarat, Maharashtra, and Karnataka, are expected to be among the most adversely affected by climate change (Sanghi, Mendelsohn, and Dinar 1998). Lonergan (1998) estimates that India's climate could become warmer under the conditions of increased atmospheric CO_2. The average temperature change is predicted to be in the range of 2.33–4.78°C, with a doubling of CO_2 concentrations. It is also likely that there will be an increase in the frequency of heavy rainfall in South and South East Asia (Watson, Zinyowera, and Moss 2001). Lal, Cubasch, Voss, *et al.* (1995) present a climate change scenario for the Indian subcontinent, taking projected emissions of greenhouse gases (GHGs) and sulphate aerosols into account. The scenario predicts an increase in annual mean maximum and minimum surface air temperatures of 0.7°C and 1.0°C over land in the 2040s with respect to the 1980s.

Methods adopted to study the impacts of climate change on crops in India are largely site-specific, based on research being carried out in farm fields (Kelkar and Bhadwal 2007). Most of the assessments have been based on changes in rainfall and temperatures. Impacts have largely been compiled for rice and wheat, which are the major staple crops grown in the country and which cover nearly 50% of the total cropped area. Some studies have further tried to translate these impacts in terms of loss in farm revenues. Table 1 provides a compilation of the various studies that have been carried out on the impacts of climate change on crops and their associated sensitivities.

Table 1 Estimates of impacts of climate change on crops in India.

Impact
Decline in yields are offset by increase in CO_2 concentrations, with the magnitude varying from one crop to the other in different regions, mostly depending on the scenario chosen.
Wheat yields in central India may drop by 2% in a pessimistic climate change scenario.
Estimates on the farm-level losses indicate that even after accounting for farm-level adaptation, a 2 °C rise in the mean temperature and a 7% increase in the mean precipitation will reduce net revenues by 12.3% in India. Loss in farm-level net revenue may range between 9% and 25% for a temperature rise of 2–3.5 °C.
Rice production in Asia could decline by 3.8% during the current century.
2 °C increase in mean air temperature could decrease rice yield by about 0.75 tonnes/hectare (t/ha) in India.
Areas suitable for growing wheat could decrease in large portions. Substantial loses are likely in rain-fed wheat.
Without the CO_2 fertilization effect, a 0.5 °C increase in winter temperature is likely to reduce wheat crop duration by one week and wheat yield by 0.45 t/ha in India. It could translate in a 10% reduction of yield from high-yielding states like Punjab, Haryana, and Uttar Pradesh.
About 2%–5% decrease in yield potential of wheat and maize for a temperature rise of 0.5–1.5 °C in India.

Continued...

Table 1 *Contd...*

Impact	Source
Wheat production would increase by about 7%–25% in 2050 if the CO_2 fertilization effects are taken into account. These are, however, limited by other conditions of nutrient application and irrigation.	
An increase of 20°C in temperature could decrease the rice yield by about 0.75 t/ha in high-yield areas. An increase of 1°C in temperature would reduce the duration of the wheat crop yield by one week, which may lead to a reduction in yield by 400–500 kilogram/hectare (kg/ha).	
Without considering the effects of CO_2 fertilization, wheat yields could decrease between 28% and 68% and, after considering the effects of CO_2 fertilization, these figures would range between +4% and –34%.	
For every 1°C rise in temperature, rice yield would decline by about 6%.	
If the temperature increases by 1°C, 2°C, and 3°C, the grain yield of wheat in Punjab will reduce by 8.1%, 18.7%, and 25.7%, respectively, and that of rice by 5.4%, 7.4%, and 25.1%, respectively.	
A doubling of CO_2 concentrations could result in a decline in wheat yields by 28%–68%.	
Under a moderate climate scenario, the salinity intrusion experienced by the coastal area of Bangladesh is resulting in serious implications on the quality of the soil in areas that were traditionally used for growing rice.	
The decline in production due to salinity intrusion could be 0.2 million tonnes (MT), which might increase to 0.56 MT under more severe scenarios.	
The production of rice and wheat could fall by 8% and 32%, respectively, by 2050.	
A 2°C increase in temperatures will result in a decrease in rice yield by about 0.75 t/ha in high-yield areas and 0.06 t/ha in low-yield coastal regions.	
There could be a decline of –2.3%–3.8% in the production of wheat in case of a 15% decrease in rainfall and 1.5%–2% increase in case of a 20% increase in rainfall, provided the irrigation is doubled during the crop growth period.	
A reduction of 40% in radiation can result in a 10% decline in wheat yields.	

Continued...

Table 1 *Contd...*

Impact	Source
An increase of 28% and 15% in yields of rice and wheat under elevated CO_2 conditions and temperature conditions of 3°C in case of wheat and 2°C in case of rice, cancels the CO_2 effect. The combined effect of CO_2 and thermal stresses might result in an increase of 21% and 4% for wheat and rice, respectively.	
In case the irrigation is affected, greater impacts are likely on the rice crop.	

Source Kelkar and Bhadwal (2007)

Crops like wheat, barley, gram, maize, and mustard are observed to be sensitive to the changes in seasonal temperatures, and their yields show a decreasing trend. Wheat is observed to be the most sensitive crop to changes in temperatures, followed by barley, maize, mustard, and gram. Variation in potential yield is observed to be large in case of regions showing relatively higher yields. All these simulations, however, assume the growth of the crop in a comparatively ideal situation, which cannot be replicated in real-world conditions. Hence, the reductions are expected to be much higher than those predicted in the analyses presented.

There are limitations to the research, given the data provided by climate change models, including its poor resolution. Studies have not taken into account the variations in input parameters (for instance, irrigation), which are known to have a significant input in enhancing agricultural production. While there are many studies that report on the nature of impacts on crops, these are largely based on specific studies that have not taken into account changes in other variables that might influence the system, assuming the changes in these variables are constant. Studies on the impacts of climate change on crop production in India are largely confined to the changes in climatic parameters and associated sensitivities in crop growth, assuming all other factors are constant, including changes in water availability, technological growth, and so on. Also, the results examined are based on field-based experiments, and therefore, chances of errors and uncertainties exist, given real-world conditions.

The results basically aim to provide an understanding of the sensitivity of crops and soil to various conditions of growth, specifying the need for more focused studies that integrate all the components and study the net effect on the crops. So far, the combined effects of CO_2 and temperature increase have been studied to some extent, and some of the results have been discussed. However, changes in technology and its impacts on crop growth have not been studied so far. For instance, crop cultivars may adapt to the changing conditions over time, and this might lead to reduced impacts on crop growth.

There are weak socio-economic linkages in the crop simulation models. These models are sensitive largely to the biophysical changes and have an inherent limitation in modelling the socio-economic linkages. Also, crop growth models have been used in isolation to study the impacts on crop productivities. The use of integrated approaches to capture the changes in other resources and their availability over time is limited. The results hardly draw any links to the pests and diseases, except for some information on the tolerance limits of different pests to rise in temperatures and changes in the relative humidity.

Agricultural losses due to climate change in India

For the translation of these impacts into economic losses, agronomic and Ricardian models have been applied in the case of India.[1] Table 2 reports the results of applying the Mendelsohn, Dinar, and Sanghi Ricardian model for India to the base and the 2080s climate variables identified in Cline (2007). The model applies seasonal monthly climate data. The results are alarming, with reductions in output potential ranging from about 30% to 35% in the southern regions to about 60% in the northern regions. The model does not include the favourable effect of carbon fertilization, but Cline considers that even after the inclusion of the effects of carbon fertilization, the losses would be severe.

---

[1] In general terms, Ricardian or cross-section models relate agricultural capacity and actual output to temperature and precipitation (and sometimes CO_2 fertilization effects) on the basis of statistical estimates from farm survey or country-level data across varying climatic zones. Agronomic model predicts changes in yield on the basis of modelling variables such as change in temperature, precipitation, and solar radiation; the effect of carbon fertilization; and crop management. Cline (2007) can be consulted for a very detailed revision of the literature on the impacts of climate change on agriculture.

Table 2 Impact of global warming by the 2080s on Indian agriculture using the Mendelsohn–Dinar–Sanghi model

Region	Percentage change	
	Net revenue	Output
North East	-91.3	-69.9
North West	-86.9	-57.9
South East	-46.9	-31.3
South West	-55.2	-36.8

Modified from Cline (2007)

Rosenzweig and Iglesias (2006) provide a query-based database that presents estimates of the impact of prospective global warming under alternative General Circulation Models (GCMs) on four major crops (wheat, rice, coarse grains, and soyabeans). The process-based dynamic crop growth models (which are essentially agronomic) incorporate the effects of change in temperature, precipitation, solar radiation, and carbon fertilization and crop management. For India, they estimate a change in yield of the order of –14.3% (average of the different models analysed). It can be observed that the Rozenweigh–Iglesias and Dinar–Sanghi results are significantly different (14.3% losses versus an average of about 50%, respectively). This shows a discrepancy in the use of these two essentially different approaches. A discussion on the limitations and advantages of the two approaches can be found in Cline (2007).

Adaptation to climate change in agriculture

Rosenzweig, Parry, Fischer, *et al.* (1993); Rosenzweig, Parry, and Fischer (1995); Rosenweig and Parry (1994); and Parry, Rosenzweig, Iglesias, *et al.* (1999) have assessed the global implications of various adaptation options on the production of grain crops using a combination of crop yield models and global trade model. The adaptation options include change in crop planting dates, increase in fertilizer use, adoption of new plant varieties, and expansion of irrigation. In the *Stern Review* (Stern, Peters, Bakhshi, *et al.* 2006), a summary of the studies addressing the benefits of adaptation in agriculture is also present, which further gets into the debate of autonomous adaptation and its suitability under certain defined conditions, including incentive mechanisms.

The benefits of adaptation are known to vary for different crops and regions and temperatures changes, but on an average, they provide approximately 10% yield benefit when compared with yields when no adaptation measure is used (Easterling, Aggarwal, Batima, *et al.* 2007). The literature on adaptation to climate change in agriculture is dominated by private adaptation, and there is little on planned adaptation to promote a general enhancement of the adaptive capacity of the agricultural system. A general finding from global-level studies is that relatively modest adaptation measures can significantly offset declines in the projected yield as a result of climate change (Agrawala and Fankhauser 2008).

The European Environmental Agency (EEA) report on Costs of Adaptation to Climate Change (EEA 2007a) presents a classification of these adaptation measures according to the time spans for adaptation in agriculture (short- and long-term adaptations) and the scale (farm level, regional level, and national level). Short-term adaptations relate to efforts to optimize production without major system changes and they are considered to be autonomous, as no other sectors are needed in their development and implementation. Long-term adaptations refer to major structural changes to overcome adversity caused by climate change and are usually the result of a planned strategy (non-autonomous or public). This classification is reproduced in Table 3. Kurukulasuriya and Rosenthal (2003) highlight the necessary supporting policies and other prerequisites and limitations of most of these adaptation options.

Table 3 Classification of adaptation options in agriculture

Short-term adaptations
Crop and livestock diversification Diversification of crop and livestock varieties, including the replacement of plant types, cultivars, hybrids, animal breeds with new varieties more tolerant to drought and heat; multi-cropping; changing crop mix.
Changes in timing of farm operations Adjusting cropping sequence and timing of irrigation
Changes in external inputs and cropping intensity Change in fertilizer and pesticide use and intensity.

Continued...

Table 3 *Contd...*
Short-term adaptations
Practices to conserve moisture Water-conserving practices include conservation through tillage and irrigation management, terracing and construction of diversions and reservoirs, reducing water use in land preparation.
Temporary (or circular) migration Includes seasonal migration where workers undertake off-farm or non-farm activities for a part of the year and return during harvest time. However, it is not clear in the literature as to what extent can climate change *per se* be attributed as the primary factor in the decision-making process of migration.
Adoption of insurance schemes Formal and informal, private and public.
Long-term adaptations
Changes in land use
Crop substitution and change in location Develop/introduce heat- and drought-resistant crop varieties; change location.
Improving water management Improving irrigation and water use efficiency and water-distribution systems.
Nutrient management Adapted not only to reflect the modified growth and yield of crops but also to changes in the turnover of nutrients in soils, including losses.
New technologies and genetic improvements
Permanent migration of labour[a]
Level of adaptations
Farm level ▪ Development of risk amelioration approaches. ▪ "Opportunistic" cropping–more effectively taking into account the environmental condition. ▪ Expand routine record keeping of weather, production, degradation, pest and diseases, and weed invasion. ▪ Tools/training to access/interpret climate data and analyse alternative management options. ▪ Learning from farmers in currently more marginal areas. ▪ Selection of varieties with appropriate thermal time, heat shock resistance, drought tolerance, high protein levels, resistance to new pests and diseases. ▪ Improve seasonal and other climate forecasting and also develop early warning systems of the likelihood of very hot days and high erosion potential. Regional level Integrate climate change in regional planning to develop optimal land use policies.

Continued...

Table 3 *Contd...*
New technologies and genetic improvements
National level ▪ *Policy* establish linkages to existing initiatives to enhance resilience. ▪ *Managing transitions* provide support during transitions to new systems. ▪ *Communication* develop industry-specific and region-specific information. ▪ R&D and training: use a participatory approach and provide the knowledge base for adaptation. ▪ Model development and application. ▪ Climate data monitoring, forecasting, and dissemination. ▪ *Breeding and selection* support programmes and ensure access to global gene pools. ▪ *Pests, diseases, and weeds* enhance quarantine measures, sentinel monitoring and management. ▪ Improve irrigation and water-use efficiency and water distribution systems. ▪ Land-use change and diversification.
[a] In the *EEA Report* on the costs of adaptation, this measure is included as an adaptation measure, although it can be taken as an impact of climate change (EEA 2007a).

Some attempts have been made to assess global estimates for the costs of the adaptation measures to climate change for developing countries. Specifically, Agrawala and Fankhauser (2008) signal the studies from the World Bank (2006), Stern (2007), Oxfam (2007), and Kelkar and Bhadwal (2007b). National-level estimates, however, are lacking in the literature.

A more detailed assessment of adaptation costs, including that in the agriculture sector, has been made by the World Bank and published recently (World Bank 2009). The estimates include what is considered to be a "planned adaptation," which results from a deliberate policy decision and excludes autonomous adaptation. The key features of this study are discussed in the introductory chapter of this book. Here, we note the estimates given in the study for the agriculture sector in India. These are derived from the World Bank study by considering the share of South Asia represented by India. The share is based on the average of the share of gross domestic product (GDP) (70%) and population (73%).[2] They imply an average

[2] Some cost items such as those of infrastructure will be more closely tied to gross domestic product (GDP), while others such as those of health will depend more on population. Yet others such as those of coastal zones will depend on both. The share of South Asian costs that have been considered here should be seen as indicative. More detailed India-specific calculations are needed, which are provided for some sectors.

annual cost of $1.3 billion per annum from 2010 to 2050. No break-up is given in the study on how the costs will vary by decade. We will compare this figure with our estimates later in the chapter.

Methodology

In this section, the methodology adopted for the estimation of the cost of adaptation for the agriculture sector in India has been described. Both autonomous and planned adaptation cost have been estimated using different methods to arrive at the results. While the estimates have been disaggregated at the level of agro-climatic zones, in some cases where data was not available, the results have been provided at the state level. In all, there are 15 classified agro-climatic zones in India (see Annexures I and II for the distribution of agro-climatic zones [ACZs] and coverage of states and districts).

Methodology for calculation of autonomous adaptation figures

Autonomous adaptation is expected to add up to a significant part of the costs of adaptation to climate change in the agricultural sector. That is why, instead of including in the other sectors of this study (such as the health and water resources sectors), autonomous adaptation is included in this chapter. However, it is not possible, due to the limited scope and means of this research project, to carry out a bottom-up estimation of the costs of the specific adaptation measures (such as diversification of crop varieties, water-conservation practices, changes in fertilizers, crop shifting, and changes in the planting timing). This approach would very likely require a number of surveys at the case study level. To overcome this limitation, we propose to take the costs of the losses on production (farmers' revenue losses) as an upper bound of the cost of autonomous adaptation. This is based on the assumption that farmers will, at maximum, expend the amount of money that they will lose due to climate change on their adaptation strategies.[3] We are aware that this approach implies an important simplification of the matter. However, we consider that the expected revenue losses of the farmers due to climate change give a

[3] Equally, one can think of these estimates as the maximum amount that the government would have to spend to prevent a loss of welfare for the farmers whose incomes declined as a result of climate change.

reasonably illustrative upper bound of what the costs of autonomous adaptation can be in the future.

For this purpose, we base our analysis on the best existing estimates on the impacts of climate change in agriculture, specifically for India, and we add value to the analysis by "breaking down" these estimates to the level of ACZs in the country by using self-estimated projections on the climatic variables, as is shown later.

Rosenzweig and Iglesias (2006) estimate the potential impacts of climate change on world food supply and generate a data set from a crop modelling study. For India, they provide yield production changes for these crops for different climatic scenarios and time horizons for different GCMs.[4]

Following the Rozenweig and Iglesias study, in our assessment, we use the equilibrium scenario with CO_2 effects at 555 parts per million (ppm) and no adaptation for the Goddard Institute for Space Studies (GISS), the Geophysical Fluid Dynamics Laboratory (GFDL), and UK Meteorological Office (UKMO); and the model outputs of HadCM3 and HadCM2, with the specifications summarized in Table 4; all models include CO_2 fertilization effects. While the equilibrium scenario is based on the assumption of a doubling of atmospheric CO_2 concentrations from pre-industrial levels (that is, from 227 ppm to 555 ppm), irrespective of the time-frame, the other scenarios from the Hadley Centre are transient in nature and are time-dependent.

Table 4 Used model output specifications considering the effects of CO_2 fertilization

Year	HadCM3	HadCM2	
		S550[a]	S750[b]
2020	475 ppm; level 1 of adaptation	410 ppm; level 1 of adaptation	424 ppm; level 1 of adaptation
2050	574 ppm; level 1 of adaptation	458 ppm; level 1 of adaptation	501 ppm; level 1 of adaptation
2080	712 ppm; level 1 of adaptation	498 ppm; level 1 of adaptation	577 ppm; level 1 of adaptation

[a] Assumes stabilization at 550 ppm in 2170; [b] Assumes stabilization at 750 ppm in 2170

[4] Data set and detailed explanation on the model are available at <http://sedac.ciesin.columbia.edu/giss_crop_study/>.

The Rosenzweig and Iglesias study was conducted over a span of eight years. The percentage crop yield losses were quantified using the average yield for 2004/05, as expressed in Equation (1).

% Loss in yield × Average yield = Loss in yield (t/ha) ... (1)

The t/ha loss in crop yields was multiplied by India's current minimum support price (MSP) of the crop (rice or wheat) in rupees to obtain the losses in revenue that would be incurred due to specified crop losses (Equation 2). The MSP has been considered here, given that it is the real "assured" amount of money that the farmer will receive, so that he can plan his adaptation strategy on this basis.

The losses in revenue were then multiplied by the total area under cultivation of the specific crop in the ACZ to arrive at the total revenue that would be lost in the entire ACZ.

Loss in yield (t/ha) × MSP of crop (Rs/tonne) = Loss in revenue (Rs/ha) ... (2)

Methodology for calculation of planned adaptation figures for 2050

This section deals with the calculation of costs of adaptation, taking into account the planned adaptation interventions that can be employed to combat the impacts of climate change. To arrive at the cost figures, a rigorous review of various plans and programmes being implemented by the Government of India was done. This led to the identification of measures, which, when implemented, have the potential to address climate-related concerns, and therefore, would help in adapting to these climate changes. The unit costs of these measures were then taken for application of these measures in areas that are likely to get affected. Measures spanning a whole range of activities from watershed development and management to creation of livelihood options were considered under various schemes and programmes, namely, the National Rural Employment Guarantee Scheme (NREGS), the National Food Security Mission (NFSM), the National Watershed Development Programme in Rain-fed Areas (NWDPRA), the Drought Prone Area Programme (DPAP), the Desert Development Programme (DDP), the Artificial

Recharge, the National Programme on Repair, Renovation and Restoration (NPRRR), and the Command Area Development and Watershed Management (CADWM).

It was assumed that the areas where rainfall in the future is expected to decrease would require adaptation measures that would contribute towards water conservation, drought-proofing techniques, and so on, and those areas that are expected to experience an increase in rainfall can employ adaptation measures such as those that would promote water conservation, building storage structures for surplus water, command area development, and so on.[5]

The following steps were executed in the process of estimation of planned adaptation. Climate-change sensitivity across regions was estimated by a review of likely increase/decrease in rainfall across different ACZs in 2050. The identification of areas that are likely to experience an increase or decrease in rainfall in the future was based on the study by Gosain, Rao, and Basuray (2004). The study provides information on the current rainfall in the major river basins of India and the future rainfall projections under a GHG emission scenario. The distribution of rainfall, by basin, that was available was grouped into ACZs by adding the quantity of rainfall (currently experienced or future projections) on the basis of the percentage of geographical area of a basin that lies in a particular ACZ. The percentage geographical area of a basin falling in an ACZ was determined by superimposing two maps delineating the ACZs of India and the river basins of India, respectively. The final output constituted the aggregate percentage geographical area of all the constitutive river basins falling within a specific ACZ.

The proportion of rainfall that a particular ACZ receives or is likely to receive was determined by multiplying the proportional areas of the river basin lying within an ACZ to the rainfall it receives or is likely to receive. This was followed by the addition of each of the values (percentage area × rainfall) of all the constitutive river basins to arrive at the final value for the ACZ. The change in rainfall for each ACZ was calculated by subtracting the rainfall that would

[5] All the measures contemplated here, except for one, concern water-related impacts. For the measure relating to the National Food Security Mission, the approach is based on the study of the temperature thresholds and crops viability.

be experienced in a GHG emission scenario in an ACZ (2050) from the rainfall that was experienced in the current year for each ACZ.[6] Formally, this can be written as

$$R_i = \sum_{i}^{j} (A_{ji} \times R_j)$$

where R_i is the total rainfall received or likely to be received in ACZ_i, j is the river basin, A_{ji} is the area of j in ACZ_i, and R_j is the total rainfall that j receives. Figure 1 indicates regions representing increase and decrease in rainfall by ACZs in 2050 under an enhanced GHG scenario.

Categorization of regions for interventions was then done on this base information that was generated. For the estimation of costs of planned adaptation, programmes and activities within were categorized into irrigation-related measures and other measures, as indicated in Table 5.

Table 5 Categorization of select programmes and activities

Irrigation-related measures		Other measures	
Irrigation-related measures in ACZs with decrease in rainfall	Increase in rainfall	All ACZs with decrease in rainfall	Temperature threshold limits
Artificial recharge	Command Area Development and Watershed Management	National Rural Employment Guarantee Scheme	National Food Security Mission
Watershed development			
DDP			
Repair, renovation, and restoration of waterbodies			
DPAP			

ACZs – agro-climatic zones; DDP – Desert Development Programme; DPAP – Drought Prone Area Propgramme

Each of these measures is considered subsequently.

[6] The approach is thus based on the expansion of current programme to areas that are likely to suffer from the change in the rain pattern, but not explicitly in the estimation of a certain level of impact and adaptation target.

ACZs with increase in rainfall	ACZs with decrease in rainfall
Western Himalayas	Central Plateau and Hills Region
Eastern Himalayas	Western Plateau and Hills Region
Lower Gangetic Plains	Southern Plateau and Hills Region
Upper Gangetic Plains	East Coast Plains and Hills Region
Trans Gangetic Plains	Western Plains and Ghat Region
Eastern Plateau and Hills	Gujarat Plains and Hills Region
	Western Dry Region

Figure 1 Increase/decrease in rainfall across different agro-climatic zones ACZs

Artificial recharge

Areas with decrease in rainfall and where gaps in irrigation were prominent were chosen for artificial recharge. Artificial recharge is a mechanism through which the water level of the groundwater is recharged to maintain the levels of the aquifers. Such an intervention is essential in areas where there is a likelihood of decline in rainfall and where huge gaps exist in the irrigation potential.

Calculation of gaps in irrigation

An attempt was made to assess the gaps in the irrigation potential based on the amount of irrigation potential already created under several schemes or projects undertaken by the government and the irrigation potential that is likely. The data was taken from the Central Water Commission, Government of India. As the source gave the information on the basis of river basins, the same method of reclassification of the data by ACZs was followed.

Gaps in irrigation (in billion cubic metres) = irrigation potential created – irrigation potential utilized

On the basis of the above analyses, five ACZs were identified where artificial recharge would improve the water situation. The Government of India has determined the area that is feasible for groundwater recharge all over the country and the state-specific recharge structures that would be used to achieve the target over the next 10 years. The information on the feasible area that can be brought under artificial recharge, by state, was obtained from the Master Plan for Artificial Recharge to Ground Water in India. Each state has a requirement of several types of recharge structures to achieve its potential in terms of artificial recharge. The unit costs of each of the recharge structures, by state, were determined on the basis of the percentage geographical area of a state that falls within a particular ACZ. Further, all the states or parts of the states that constitute an ACZ were grouped, and the corresponding numbers of structures that would be required to recharge the total area of the ACZ were determined.

% area of state$_1$ lying within ACZ$_1$ × number of recharge structures required for state$_1$ = number of recharge structures required

An ACZ may have whole or parts of several states lying within its boundaries. A summation of all the portions of recharge structures being used in each of the states that lie in a specific ACZ gave the total number of each type of recharge structures that would be required by a particular ACZ.

[(Total number of recharge structure$_1$ × unit cost of recharge structure$_1$) + (total number of recharge structure$_2$ × unit cost of recharge structure$_2$) + … (total number of recharge structure$_n$ × unit cost of recharge structure $_n$)]$_{\text{state1}}$ + [(total number of recharge structure$_1$ × unit cost of recharge structure$_1$) + (total number of recharge structure$_2$ × unit cost of recharge structure$_2$) + … (total number of recharge structure$_n$ × unit cost of recharge structure$_n$)]$_{\text{state2 … etc.}}$ = total cost of recharge structures for the ACZ

Repair, renovation, and restoration of waterbodies

Repair, renovation, and restoration (RRR) of waterbodies would prove to be an effective measure for adaptation in areas that are likely to experience a decrease in rainfall in the future. On the basis of the previous assessments, the ACZs and the states that are covered within those ACZs were sorted. The estimated costs to restore, repair, and renovate select waterbodies in these states were obtained, and the unit costs were arrived at.

Unit cost of repairing/restoring/renovating a waterbody in a district = [total estimated cost/total number of waterbodies selected in the district]

Given that the costs would vary by the type as well as the scale of activity undertaken, a range of costs was considered. The lowest unit cost and the highest unit cost for RRR were then used to arrive at the lowest and highest costs to undertake RRR activities for the total number of waterbodies that had been selected. The final estimate thus provided is an upper and a lower bound of the total costs that would be incurred in case RRR measures have to be undertaken.

Lower bound = lowest cost of repairing/restoring/renovating × total number of waterbodies selected in the area that would experience a decrease in rainfall in the future

Upper bound = highest cost of repairing/restoring/renovating × total number of waterbodies selected in the area that would experience a decrease in rainfall in the future

Desert Development Programme

The programme is being implemented in a select few states and does not have complete coverage. The programme is being implemented through projects. The total number of projects being implemented in each state was obtained. These projects were found to be concentrated in areas that are projected to further experience a decline in rainfall under an enhanced climatic situation. To arrive at the total cost that would be incurred for the implementation of this programme in the rainfall-deficit areas, the number of projects was multiplied to the unit cost for the completion of a project, which was reported to be approximately Rs 3 million (MoRD 2001).[7]

Number of Desert Development Programme (DDP) projects in states that are covered under DDP and would experience a decrease in rainfall in the future × Rs 3 million = total cost of implementing DDP in areas

National Food Security Mission

On the basis of the crops predominantly grown in the area, several districts to be brought under the National Food Security Mission (NFSM) have been identified all over India. In the following assessment, two measures operational under the scheme of NFSM rice have been considered. These are as follows.

- Support for the promotion of hybrid rice seeds and their production and distribution.
- Distribution of sprinkler sets.

---

[7] Details also available at <http://sirsa.gov.in/htfiles/ddp.pdf>.

It is expected that due to climate change, certain areas will exceed the threshold limits, both higher and lower for growth of certain crops, including rice. Therefore, it is required that hybrid varieties be promoted that are resistant to such type of changes, for instance, increase/decrease in temperatures and changes in rainfall and other associated parameters. The unit cost for implementation of the abovesaid measure was reported to be Rs 6000 per quintal. In the case of rice, temperatures conducive to the growth of the crop lie between 15°C and 30°C (Venkataraman and Krishnan 1992), though it is known to vary depending on the variety used and other local factors. The projected data for minimum and maximum temperatures for the ACZs was downscaled using the Regional Climate Model (RCM) outputs for the time period 2071–97 for the A2 scenario.[8] It was felt that those ACZs where temperature thresholds are likely to be crossed under an enhanced climate scenario should be targeted for implementation of the measure. ACZs exceeding threshold limits beyond tolerance for the rice crop were thus identified for intervention of the measure (Table 6).

> Loss in area (ACZ) = f(*x*) exceedance in threshold limits of the crops (based on temperature tolerance)

Table 6 Agro-climatic zones with temperatures beyond threshold limits

Region	T_{min}[a]	T_{max}
Western Himalayan	-13.16	1.18
Eastern Himalayan	2.96	14.61
Lower Gangetic Plains	13.55	25.32
Middle Gangetic Plains	13.80	26.98
Upper Gangetic Plains	3.98	18.43
Trans-Gangetic Plains	7.51	23.00
Eastern Plateau and Hills	17.23	29.88
Central Plateau and Hills	15.17	30.00
Western Plateau and Hills	19.63	32.78
Southern Plateau and Hills	21.48	32.58

Continued...

[8] A2 scenario is the Intergovernmental Panel on Climate Change (IPCC) Scenario that is one of the most extreme scenarios towards the higher end of the emissions spectrum.

Table 6 *Contd...*

Region	T_{min}[a]	T_{max}
East Coast Plains and Hills	19.39	31.04
Western Plains and Ghat	21.16	33.13
Gujarat Plains and Hills	17.21	32.25
Western Dry	12.59	28.88

ACZs - agro-climatic zones
[a]Cells highlighted are those for which the threshold levels have been found to have exceeded for crop growth.

The total production of rice for 2005/06 (in quintals) was obtained for these ACZs. In case the seeds are to be replaced with advanced varieties likely to be tolerant to increase in temperatures, the unit cost for promotion and distribution of hybrid seeds is applied to arrive at the total costs. It is assumed that 50% of the production is sown back into the farm and, therefore, the costs have been computed accordingly.

Fifty per cent of the rice production in $ACZ_I \times 6000$ = total cost of implementation of measure in ACZ_I

The distribution of sprinkler sets was targeted for ACZs that are likely to experience a decrease in rainfall in the future. The total area under cultivation of rice and wheat in each of the identified ACZs was multiplied to the unit cost of Rs 15 000/ha.

(Area under cultivation of rice in the ACZ + area under cultivation of wheat in the ACZ) × 15 000 = total cost of distributing sprinkler sets in the cultivated area of the ACZ

National Watershed Development Programme in Rain-fed Areas

The ACZs that would experience a decrease in rainfall in the future were considered for the development of watersheds. The total area in the states that are already covered by the programme in these ACZs was subtracted from the total area of these ACZs to arrive at the area where the development of watersheds is not being promoted presently but would require such interventions so as to avoid the detrimental impacts of climate change on agriculture.

> Total area of ACZs that are likely to experience a decrease in rainfall in the coming years – cumulative area under the National Watershed Development Programme in Rain-fed Areas (NWDPRA) at present = total area that requires watershed development (ha)

To arrive at the total cost for development of watersheds in relevant areas, the total area in the identified ACZs, which is presently not targeted, was multiplied by Rs 15 000, which is the unit cost for the development of watersheds. However, the costs are known to vary in their application from region to region, depending on the scales of application and the feasibility and the nature of activities that are required to be undertaken.

> Total area that requires watershed development (ha) × 15 000 = total cost of development of watersheds

The total cost was then divided into administrative costs, community organization costs, training costs, and cost of works on the basis of the Integrated Wasteland Development Programme (IWDP) distribution of costs.

> Administration costs = 10% × total cost of development of watersheds
> Community organization cost = 5% × total cost of development of watersheds
> Training costs = 5% × total cost of development of watersheds
> Works costs = 80% × total cost of development of watersheds

Drought Prone Area Programme

To arrive at the cost estimate, the total area of all ACZs that are likely to experience a decrease in rainfall was multiplied by Rs 6000, which is the cost of treatment of one hectare of land area.

> Total area of ACZs that are likely to experience a decrease in rainfall in the future × Rs 6000 = total cost of implementing the programme

National Rural Employment Guarantee Scheme

Only those states that are covered within these ACZs and that are likely to experience a decrease in rainfall in the future were considered. The population projections for 2030 were obtained from the assessments based on the in-house population growth rate for each state based on their individual growth rates for the period 2001–08. The population of each state was projected for 2030. The population figures were translated into the number of households based on the assumption that an average household consists of five members. The unit cost of implementation of the scheme was calculated by dividing the total number of households that were provided employment in 2008 with the total amount that was spent on the scheme to achieve its target in 2008. Of the total population projections for 2030, by state, the population in the rural areas was determined by using the urban population proportion for India in 2030, as estimated by Jiang and O'Niell (2009).[9] The study estimates that the proportion of urban population in India in 2030 would be 38%. The rural population of India was arrived at by subtracting the urban population figure from the total population figure of 2030. To arrive at the total number of rural households that would be in existence in 2030 in India, the population figure was divided by five. The total cost of implementing the programme in 2030 in all the rural households of India was calculated by multiplying the unit cost of implementation of the programme with the total number of rural households in 2030.

Command Area Development and Watershed Management Methodology

Command area development would be beneficial in areas that would experience an increase in rainfall in the future. The total minor irrigation potential for each state was obtained. On the basis of the minor irrigation coverage under the plan periods (up to the Ninth Five-year plan), rest of the area was identified for expanding the irrigation potential. The unit cost for achieving the minor irrigation potential was then used to arrive at the total costs. The information on the ultimate minor irrigation potential of the states was obtained from the Ministry of Water Resources, Government of India.

---

[9] Details available at <http://www.populationaction.org/Issues/Population_and_Climate_Change/IR-09-026.pdf?>.

Ultimate irrigation potential – cumulative potential created till the end of the Ninth Five-year Plan = additional area that needs to be brought under the programme (ha)

The total additional area that needs to be brought under the programme when multiplied with the cost for treatment of one hectare of land gave us the additional expenditure that needs to be made.

Additional area that needs to be brought under the programme – unit cost of implementation of programme (Rs 15 000) = total expenditure required (Rs)

The costs were then grouped into ACZs on the basis of the percentage of area of constituting states that fall within its boundaries.

Results

Table 7 summarizes the results for the autonomous adaptation aggregated across all the regions and for the different models and scenarios used.

Results for the different scenarios are presented in Annexure III as per ACZs outlined in Annexures I and II. In Table 7, we present the total revenue losses predicted for the entire India according to different models and time horizons.

Table 7 Predicted revenue losses (million rupees a year) for the entire India

Scenarios	Rice			Wheat		
	2020	2050	2080	2020	2050	2080
HadCM2 S750	33 274	70 958	78 174	22 510	31 514	36 016
HadCM2 S550	27 728	55 189	78 174	18 008	31 514	31 514
HadCM3	44 365	55 189	104 232	27 012	27 012	49 522

Source Modified using Rosenzweig and Iglesias (2006) predicted yield changes

These estimates indicate an annual loss of about $1 billion to $1.5 billion by 2020 and increasing up to $1.8 to $2.2 billion by 2050, using an exchange rate of 1$ = Rs 46.86. Table 8 summarizes the costs and expenditures for the planned adaptation measures estimated in this study, per ACZ.

Table 8 Summary of the costs of select options for planned adaptation by agro-climatic zones (costs in million rupees)

ACZs	Artificial recharge to ground-water	RRR	NFSM: promo-tion and distribu-tion of hybrid seeds	NFSM: distri-bution of sprinkler sets	DDP	NWDPRA	DPAP	NREGS	CADWM	Total
Western Himalayan			37767						56	37 823
Eastern Himalayan			198 734							198 734
Lower Gangetic Plains									33	33
Middle Gangetic Plains									38	38
Upper Gangetic Plains									32	32
Trans Gangetic Plains										0
Eastern Plateau and Hills									3	3
Central Plateau and Hills	11 647			80 317		111 626	236 411	56 589	117	496 706
Western Plateau and Hills	28 315		98 451	38 127		141 648	201 752	78 710		587 003
Southern Plateau and Hills	24 615			48 701		121 823	236 915	64 657		496 711
East Coast Plains and Hills				52 133		77 121		44 880		174 134
Western Plains and Ghat			33 831							33 831
Gujarat Plains and Hills	5206		39 762	20 892		26 822	118 016	32 665		243 363
Western Dry	5916			5758		64 398	101 619	21 373		199 064
Island										0
Total for India	75 699	843-20286	408 545	245 928	33 825	543 438	894 713	298 874	279	2 521 587

ACZs – agro-climatic zones; RRR – repair, renovation, and restoration; NFSM – National Food Security Mission; DDP – Desert Development Programme; NWDPRA – National Watershed Development Programme in Areas; DPAP – Drought Prone Area Programme; NREGS – National Rural Employment Guarantee Scheme; CADWM – Command Area Development and Watershed Management

Note Data in the blank grids of the table is not applicable.

The amount is a cumulative amount required for the implementation of these measures, given the scales that are required to be addressed when impacts are taken into account in 2050. This can be spread in terms of costs required over a 40-year term, from 2010 to 2050. Taking into account the outlined measures, the costs for adaptation would be about $53.8 billion. On the basis of its implementation over the next 40 years, the annual costs could turn out to be $1.3 billion/year. However, individually, the impacts on crops have not been covered, and the overall requirement for R&D in agriculture has been excluded, which actually ought to be a major investment on the scale of millions of rupees each year.

Conclusion

In this chapter, we have estimated the costs of adaptation to climate change in India for the agricultural sector. Contrary to the other sectors, autonomous adaptation is included, as it is expected to add a significant part to the total costs.

For autonomous adaptation, an upper bound has been established by assuming that farmers will at the most expend the amount of money that they will lose due to climate change. We are aware that this approach implies an important simplification of the matter. Several assumptions need to be made, not only regarding climatic projection but also production losses, CO_2 fertilization effects, neglecting certain levels of adaptation, and so on. The other important assumption is that the farmers will behave according to losses that will occur in the future and that they might not be aware of, or be able to predict, the same. However, we consider that the expected revenue losses to farmers due to climate change give a reasonably illustrative upper bound of what the costs of autonomous adaptation can be in the future.

Our estimates indicate an annual loss of about $1 billion to $1.5 billion by 2020, increasing from $1.8 billion to $2.2 billion for autonomous adaptation by 2050.

For planned adaptation, we have used the information of existing ongoing programmes, and we have calculated the additional costs of extending these programmes to areas in which the rainfall pattern is expected to change (for the Food Security Mission, we used the temperature threshold exceedance and crops viability for identifying vulnerable areas in the future). Our estimates add up to $53.3 billion to $53.8 billion cumulative, taking into account the impacts for 2050. On the basis of its implementation over the next 40 years, the annual costs could turn out to be about $1.3 billion/year.

These two estimates of the costs of adaptation are, of course, not additive. They represent different approaches to arriving at a figure, which captures a complex reality of adjusting to climate change. The first estimate of $1.8 billion to $2.2 billion by 2050 is based on the estimated losses in agriculture from climate change if no action is taken. Arguably these are an upper bound of what individuals and governments would need to spend to prevent a loss in the welfare of farmers relative to the baseline case of no climate change. In practice, one would expect to be able to adapt at a lower cost than that, and the second approach indicates this is the case. The measures needed to make good the losses of resources due to climate change amount to about $1.3 billion/year. We have noted, however, that these do not include all possible and justified measures (such as additional R&D on improved crops, for example). Furthermore, we are not able to guarantee that they will restore the *status quo ex ante* without climate change even if all these measures are taken (although we believe that they would go a long way towards meeting that objective). Thus, the "true" annual costs may lie somewhere above the $1.4 billion and below the $1.8–2.2 billion figures.

It is valuable to compare our estimates with those of others, which have been derived for adaptation to agriculture in India. As noted earlier, the closest we have is the World Bank, 2010 study, which gives figures for South Asia, and from which we have derived an estimate of around $1.3 billion a year from 2010 to 2050. It may be coincidental, but the estimate from that study and this one are remarkably close, although both have many gaps and use quite different approaches.

In addition to providing an estimate of the costs at the national level, our study has a number of other important results. It is, for example, the first to downscale the information at the ACZ level, although further progress can still be made in the area of automomous adaptation, by using climatic projection at the regional level.

While there have been other studies that have attempted to compute farm-level losses, the losses have been presented in terms of percentage decline in revenues, and the absolute numbers in terms of the costs are not well highlighted. This study tries to fill this gap by presenting the results in terms of the costs, both for autonomous and planned adaptation. However, to be able to do this, we were dependent on the limited number of studies that were available on impacts and productivity losses, on the basis of which the costs were computed. Coarse resolution data was used in many cases, and therefore, the results are broad-based, though indicative of the kind of losses that are expected. For instance, the Rosenweig and Iglesias study, on the basis of which results were obtained, cover, in all, only two stations in India. Integrated assessments were not carried out, and therefore, a holistic sense of the kind of costs that will be incurred, given the losses and the interventions proposed, is still to be arrived.

ANNEXURES

Annexure I Agro-climatic zones in India and their classification

Agro-climatic zones	State	Districts
Western Himalayan Zone	Jammu and Kashmir	Jammu, Baramula, Anantnag, Pulwama, Srinagar, Badgara, Kathua, Kargil, Ladakh, Kupwara, Punch, Rajauri and Udhampur, Doda
	Uttaranchal	Uttarkashi, Dehra Dun, Haridwar, Chamoli, Tehri Garhwal, Pauri Garhwal, Almora, Nainital, Udham Singh Nagar, Pithoragarh, Champavath, Bageshwar
	Himachal Pradesh	Lahul and Spiti, Kinnaur, Bilaspur, Chamba, Hamirpur, Kangra, Kullu, Mandi, Shimla, Sirmaur, Sola, Una
Eastern Himalayan Region	Arunachal Pradesh, Meghalaya, , Manipur, Mizoram Tripura Nagaland, Sikkim	All districts
	Assam	Lakhimpur, Sibsagar, Cachar, Dibrugarh, Jorhat, Karimganj, Karbi Anglong, North Cachar Hills, Barpeta, Dhubri, Nagaon, Darrang, Kamrup, Goalpara, Kokrajhar and Sonitpur
	West Bengal	Jalpaiguri, Kooch Behar, Darjeeling
Lower Gangetic Plains	West Bengal	All districts of West Bengal (Bankura, Birbhum, Midnapur) except Jalpaiguri, Kooch Behar, Darjeeling, and Purulia
Middle Gangetic Plains Region	Bihar	Paschim Champaran, Purbi Champaran, Gopalganj, Siwan, Saran, Sitamarhi, Muzzaffarpur, Vaishali, Madhubani, Darbhanga, and Samastipur. Madhepura, Purnia, Katihar, Khagria, Begusarai and Saharsha, Bhojpur, Rohtas, Patna, Aurangabad, Jehanabad, Nalanda, Gaya, Nawada, Munger and Bhagalpur, Bhabua, Jamui
	Eastern Uttar Pradesh	Baharaich, Gonda, Balrampur, Basti, Gorakhpur, Sidharth Nagar, Maharajgunj, Kushinagar, Deoria, Azamgarh, Mau, Balia, Faizabad, Ghazipur, Jaunpur, Sant Ravidas Nagar, Varanasi, Mirzapur and Sonbhadra, Shravasti

Continued...

Annexure I *Contd...*

Agro-climatic zones	State	Districts
Upper Gangetic Plains	Uttar Pradesh (Western and Central)	Allahabad, Fatehpur, Pratapgarh, Sultanpur, Rae Bareili, Unnao, Lucknow, Bara Banki, Sitapur, Hardoi, Kheri, Pilibhit, Shahjahanpur, Bareilly, Rampur, Moradabad, Bijnor, Saharanpur, Muzaffarnagar, Meerut, Baghpat, Ghaziabad, Bulandshahr, Badaun, Aligarh, Mathura, Agra, Etah, Farrukhabad, Kannauj, Mainpuri, Firozabad, Etawah, Kanpur Dehat and Kanpur.
Trans-Gangetic Plains Region	Chandigarh	Chandigarh
	Delhi	All districts
	Haryana	Kurukshetra, Karnal, Jind, Sonipat, Rohtak, Faridabad, Gurgaon, Mahendragarh, Bhiwani, Hisar and Sirsa, Fatehabad, Jhajjar, Rewari
	Punjab	Amritsar, Kapurthala, Jalandhar, Ludhiana, Patiala, Firozpur, Faridkot, Bhatinda and Sangrur, Hoshiarpur, Gurdaspur, Rupnagar and Ambala
	Rajasthan	Ganganagar
Eastern Plateau and Hills Region	Chhattisgarh	Durg, Bastar, Bilaspur,
	Jharkhand	Palamu, Gurala, Lohardaga, Ranchi and Singhbhum, Godda, Sahibganj, Dumka, Bokaro, Garhwal, Koderma, Chatra, Pakur, Latehar, Jamatra Deogarh, Dhanbad, Giridih, Hazaribag and Palamu.
	Madhya Pradesh	Balaghat, Durg, Rajnandgaon, Raipur and Bilaspur Bastar Shahdol, Surguja and Raigarh
	Maharashtra	Bhandara, Chandrapur, and Gadchiroli
	Orissa	Sundargarh, Kendujhar and Mayurbhanj, Koraput, Kalahandi and Phulabani
	West Bengal	Purulia

Continued...

Annexure I *Contd...*

Agro-climatic zones	State	Districts
Central Plateau and Hills Region	Madhya Pradesh	Datia, Chattarpur, Tikamgarh, Mandla, Sidhi, Rewa, Satna, Panna, Jabalpur, Seoni, Sehore, Raisen, Bhopal, Vidisha, Damoh, Sagar, Betul, Chhindwara, Narsimhapur, Hoshangabad, Morena, Bhind, Gwalior, Shivpuri, Guna
	Rajasthan	Dungarpur, Banswara, Sirohi, Pali, Bhilwara, Udaipur, Chittaurgarh, Bundi, Kota, Ajmer, Tonk, Jaipur, Alwar, Bharatpur, Sawai Madhopur, Dholpur
	Uttar Pradesh	Jalaun, Jhansi, Lalitpur, Hamirpur, Banda, Chitrakut, Mahoba
Western Plateau and Hills Region	Madhya Pradesh	West Nimar, East Nimar, Dewas, Indore, Dhar, Ratlam, Ujjain, Shajapur, Rajgarh, Jhalawar and Mandsaur
	Maharashtra	Kolhapur, Satara, Pune, Nashik and Jhabua, Ahmednagar, Dhule, Sangli, Solapur, Osmanabad, Latur, Bid, Aurangabad, Jalna, Parbhani, Akola, Amravati, Buldana, Jalgaon, Nagpur, Wardha, Yavatmal and Nanded
	Rajasthan	Jhalawar
Southern Plateau and Hills Region	Andhra Pradesh	Chitoor, Cuddapah, Kurnool, Anantapur, Rangareddy, Mahbubnagar, Nalgonda, Hyderabad, Adilabad, Nizamabad, Karimnagar, Medak, Warangal and Khammam
	Karnataka	Bidar, Gulbarga, Bijapur, Belgaum, Dharwad, Raichur, Bellary, Chitradurga, Tumkur, Bangalore, Hassan, Mandya, Mysore
	Tamil Nadu	Dharmapuri, Salem, Periyar, Coimbatore, Anna, Madurai, Tiruchchirappalli, and Pudukkottai
East Coast Plains and Hills Region	Tamil Nadu	Chengalpattu/Madras, North Arcot, South Arcot, Thanjavur, Kamarajar, PM Ramanathapuram, Ramanathapuram, Tuticorin, and Tirunelveli

Continued...

Annexure I *Contd...*

Agro-climatic zones	State	Districts
	Andhra Pradesh	Srikakulam, Vizianagaram, Vishakhapatnam, East Godavari, West Godavari, Krishna, Guntur, Prakasam, and Nellore
	Orissa	Baleshwar, Cuttack, Puri and Ganjam
	Puducherry	Karaikal, Mahe, Puducherry, and Yanam
Western Plains and Ghat Region	Goa	All districts
	Karnataka	Uttara Kannada, Dakshin Kannad, Shimoga, Chikmagalur, and Kodagu
	Kerala	Thiruvananthapuram, Kollam, Alappuzha, Ernakulam, Thrissur, Malappuram, Kozhikode, Kannur, Kasaragod, Kottayam, Pathanamthitta, Palakkad, Idukki, and Wayanad
	Maharashtra	Thane, Greater Bombay, Raigarh, Ratnagiri, and Sindhudurg
	Tamil Nadu	Kanyakumari and Nilgiri
Gujarat Plains and Hills Region	Gujarat	All
	Dadra and Nagar Haveli	All
	Daman and Diu	Daman and Diu
Western Dry Region	Rajasthan	Bikaner, Barmer, Jaisalmer, Churu, Jhunjhunun, Nagaur, Sikar, Jodhpur, and Jalore
Island region	Andaman and Nicobar	North and middle Andaman, South Andaman and Nicobar
	Lakshadweep	Lakshadweep

Annexure II

Demarcation of agro-climatic zones in india

1. Western Himalayan Region
2. Eastern Himalayan Region
3. Lower Gangetic Plains Region
4. Middle Gangetic Plains Region
5. Upper Gangetic Plains Region
6. Trans-Gangetic Plains Region
7. Eastern Plateau and Hills Region
8. Central Plateau and Hills Region
9. Western Plateau and Hills Region
10. Southern Plateau and Hills Region
11. East Coast Plains and Hills Region
12. West Coast Plains and Ghat Region
13. Gujarat Plains and Hills Region
14. Western Dry Region
15. The Islands Region

Annexure III

Results per agro-climatic zone (ACZ) and measure

Table I Revenue loss estimates based on Goddard Institute for Space Studies (GISS), Geophysical Fluid Dynamics Laboratory (GFDL), and UK Meteorological Office (UKMO) projections

ACZs	Revenue losses (in million rupees)					
GCMs	GISS		GFDL		UKMO	
Crops	Rice	Wheat	Rice	Wheat	Rice	Wheat
Western Himalayan Region	800	2334	1101	3034	2002	5134
Eastern Himalayan Region	5408	177	7436	230	13521	389
Lower Gangetic Plains Region	18896	1673	25982	2175	47241	3681
Middle Gangetic Plains Region	3137	6383	4314	8299	7844	14044
Trans Gangetic Plains Region	17969	49455	24708	64291	44924	108801
Eastern Plateau and Hills Region	14841	389	20406	505	37102	855
Central Plateau and Hills Region	1183	16847	1627	21901	2959	37063
Western Plateau and Hills Region	1115.712	6797.6874	1534.104	8836.994	2789.28084	14954.9123
Southern Plateau and Hills Region	10781.94	421.44312	14825.17	547.8761	26954.8464	927.174864
East Coast Plains and Hills Region	13015.69	0.04066	17896.57	0.052858	32539.2272	0.089452
Western Plains and Ghat Region	NA	0.21836	NA	0.283868	NA	0.480392
Gujarat Plains and Hills Region	1577.154	3598.976	2168.587	4678.669	3942.88512	7917.7472

Table II Revenue loss estimates based on HadCM3 projections

ACZs	Revenue losses (in million rupees)					
	2020		2050		2080	
	Rice	Wheat	Rice	Wheat	Rice	Wheat
Western Himalayan Region	400.4586	700.2139	498.1608	700.2139	940.8365	1283.725
Eastern Himalayan Region	2704.276	53.12696	3364.055	53.12696	6353.421	97.39943
Lower Gangetic Plains Region	9448.277	501.9827	11753.43	501.9827	22197.76	920.3017
Middle Gangetic Plains Region	1568.951	1915.159	1951.737	1915.159	3686.09	3511.125
Trans Gangetic Plains Region	8984.993	14836.52	11177.11	14836.52	21109.32	27200.29
Eastern Plateau and Hills Region	7420.535	116.7081	9230.967	116.7081	17433.79	213.9648
Central Plateau and Hills Region	591.8401	5054.153	736.2349	5054.153	1390.468	9265.947
Western Plateau and Hills Region	557.8562	2039.306	693.9596	2039.306	1310.626	3738.728
Southern Plateau and Hills Region	5390.969	126.4329	6706.236	126.4329	12665.53	231.7937
East Coast Plains and Hills Region	6507.845	0.012198	8095.603	0.012198	15289.52	0.022363
Western Plains and Ghat Region	NA	0.065508	NA	0.065508	NA	0.120098
Gujarat Plains and Hills Region	788.577	1079.693	980.9708	1079.693	1852.681	1979.437

Table III Gaps in irrigation potential created and utilized

Agro-climatic zones	Gap in irrigation potential created and utilized (billion cubic metres)
Western Himalayan Region	37.1488
Eastern Himalayan Region	573.9200
Lower Gangetic Plains Region	37.1488
Middle Gangetic Plains Region	92.8720
Upper Gangetic Plains Region	92.8720
Trans Gangetic Plains Region	13.9308
Eastern Plateau and Hills Region	73.5165
Central Plateau and Hills Region	158.2026
Western Plateau and Hills Region	122.6509
Southern Plateau and Hills Region	30.2585
East Coast Plains and Hills Region	3.9550
Gujarat Plains and Hills Region	14.5526
Western Dry Region	7.7066

Table IV Artificial recharge to groundwater

Agro-climatic zones	Cost (in million rupees)
Central Plateau and Hills Region	11 647.000
Western Plateau and Hills Region	28 315.490
Southern Plateau and Hills Region	24 615.420
Gujarat Plains and Hills Region	5 206.500
Western Dry Region	5 916.888

Table V Cost estimates for rooftop rainwater harvesting

Agro-climatic zones	Cost (in million rupees)
Western Himalayan Region	2600.00
Eastern Himalayan Region	382.50
Lower Gangetic Plains Region	62.25
Middle Gangetic Plains Region	2370.00
Upper Gangetic Plains Region	2340.00
Trans Gangetic Plains Region	3364.00
Eastern Plateau and Hills Region	3476.05

Continued...

Table V *Contd...*

Agro-climatic zones	Cost (in million rupees)
Central Plateau and Hills Region	3553.2
Western Plateau and Hills Region	6460.8
Southern Plateau and Hills Region	6460.8
East Coast Plains and Hills Region	4162.0
Western Plains and Ghat Region	1815.7
Gujarat Plains and Hills Region	4660.0
Western Dry Region	2200.0
Island Region	230.0

The cost of RRR of waterbodies ranges from Rs 843.473 million to Rs 20 286 million. The implementation of the DDP would require an additional expenditure of Rs 33 825 million.

Table VI Estimations of expenditure required for promotion and distribution of hybrid rice seeds

Agro-climatic zones	Cost (in million rupees)
Western Himalayan Region	37 767
Eastern Himalayan Region	198 734
Western Plateau and Hills Region	98 451
Western Plains and Ghat Region	33 831
Gujarat Plains and Hills Region	39 762

Table VII Estimations of expenditure required for distribution of sprinkler sets

Agro-climatic zones	Cost (in million rupees)
Central Plateau and Hills Region	80 317.785
Western Plateau and Hills Region	38 127.315
Southern Plateau and Hills Region	48 701.610
East Coast Plains and Hills Region	52 133.115
Gujarat Plains and Hills Region	20 892.000
Western Dry Region	5 758.305

Table VIII Cost estimates for watershed development	
Agro-climatic zones	Cost (in million rupees)
Central Plateau and Hills Region	111 626.10
Western Plateau and Hills Region	141 648.70
Southern Plateau and Hills Region	121 823.30
East Coast Plains and Hills Region	77 121.19
Gujarat Plains and Hills Region	26 822.71
Western Dry Region	64 398.27

Table IX Costs estimates for the implementation of Drought Prone Area Programme	
Agro-climatic zones	Cost (in million rupees)
Central Plateau and Hills Region	236 411.76
Western Plateau and Hills Region	201 752.40
Southern Plateau and Hills Region	236 915.14
East Coast Plains and Hills Region	118 423.25
Gujarat Plains and Hills Region	118 016.40
Western Dry Region	101 619.24

Table X Cost estimates for the implementation of National Rural Employment Guarantee Scheme (NREGS)	
Agro-climatic zones	Cost (in million rupees)
Central Plateau and Hills Region	56588.823
Western Plateau and Hills Region	78710.459
Southern Plateau and Hills Region	64657.613
East Coast Plains and Hills Region	44880.436
Gujarat Plains and Hills Region	32665.837
Western Dry Region	21373.340

Table XI Cost estimates for Command Area Development and Watershed Management (CADWM)	
Agro-climatic zones	Cost (in million rupees)
Western Himalayan Region	56.852760
Lower Gangetic Plains Region	33.912800
Middle Gangetic Plains Region	38.921480
Upper Gangetic Plains Region	32.913090
Trans Gangetic Plains Region	3.200134
Eastern Plateau and Hills Region	117.264400

References

Aggarwal P K and Mall R K. 2002. **Climate change and rice yields in diverse agro-environments of India. II. Effect of uncertainties in scenarios and crop models on impact assessment.** *Climatic Change* **52**: 331–343

Aggarawal P K and Sinha S K. 1993. **Effect of probable increase in carbon dioxide and temperature on wheat yields in India.** *Journal of Agricultural Meteorology* **48**(5): 811–814

Agrawala S and Fankhauser S (eds). 2008. ***Economic Aspects of Adaptation to Climate Change.*** Paris: OECD Publishing. p. 133.

Cline W R. 2007. ***Global Warming and Agriculture: impact estimates by country.*** Washington, DC: Center for Global Development and the Peterson Institute for International Economics.

Dinar A, Mendelsohn R, Evenson R, Parikh J, Sanghi A, McKinsey J. 1998. ***Measuring the Impact of Climate Change on Indian Agriculture.*** Washington, DC: The World Bank

Easterling W E, Aggarwal P K, Batima P, Brander K M, Erda L, Howden S M, Kirilenko A, Morton J,. Soussana J F, Schmidhuber J, Tubiello F N. 2007. **Food, fibre, and forest products.** In *Climate Change 2007: Impacts, Adaptation and Vulnerability. Contribution of Working Group II to the Fourth Assessment Report of the Intergovernmental Panel on Climate Change*, pp. 273–313, edited by M L Parry, O F Canziani, J P Palutikof, P J van der Linden, and C E Hanson. Cambridge, UK: Cambridge University Press

EEA (European Environment Agency). 2007a. **Costs of adaptation to climate change: a review of assessment studies with a focus on methodologies used.** [Agreement No 3602/B2005, EEA under the Framework Contract No EEA/AIR/04/004, final working paper]

EEA (European Environment Agency). 2007b. ***Climate Change: the cost of inaction and the cost of adaptation.*** Luxembourg: EEA. [EEA Technical Report No. 13/2007.]

GoI (Government of India). 2001. ***National Action Programme to Combat Desertification 2001: status of desertification,*** *vol. I.* New Delhi: Ministry of Environment and Forests. GoI. 294 pp.

GoI (Government of India). 2004. ***India's Initial National Communication to the United Nations Framework Convention on Climate Change.*** New Delhi: Ministry of Environment and Forests, Government of India

GoI (Government of India). 2008. **Agriculture, rural development, industry, services, and physical infrastructure.** In *Eleventh Five Year Plan (2007–2012),* Vol. III. New Delhi: Oxford University Press.

Gosain A K, Rao S, and Basuray D. 2004. **Climate change impact assessment on hydrology of Indian river basins.** *Current Science* **90**(3): 346–353

IPCC (Intergovernmental Panel on Climate Change). 2007. ***WG II Report on Impacts, Vulnerability and Adaptation.*** Cambridge: Cambridge University Press

Jiang L and O'Neill B C. 2009. ***Household Projections for Rural and Urban Areas of Major Regions of the World.*** Laxenburg, Austria: IIASA (International Institute for Applied Systems Analysis). [IIASA Interim Report IR-09-026]

Kelkar U and Bhadwal S. 2007. **South Asian Regional Study on Climate Change Impacts and adaptation: implications for human development.** [Human Development Report Office; Occasional Paper]. 44 pp. Details available at <http://hdr.undp.org/en/reports/global/hdr20072008/papers/kelkar_ulka%20and% 20bhadwal_suruchi.pdf>

Kurukulasuriya and Rosenthal. 2003. ***Climate Change and Agriculture: a review of impacts and adaptation***. Washington, DC: The World Bank Environment Department. [Climate change series paper. Paper 91]

Kumar K S Kavi and Parikh J. 1998. **Climate change impacts on Indian agriculture: the Ricardian approach.** In *Measuring the Impact of Climate Change on Indian Agriculture*, edited by A Dinar, R Mendelsohn, R Everson, J Parika, A Sanghi, K Kumar, J Mckinsey and S Lonergan. Washington, DC: The World Bank [World Bank Technical Paper No 402]

Lal M, Cubasch U, Voss R, Waszkewitz J. 1995. **Effect of transient increases in greenhouse gases and sulphate aerosols on monsoon climate.** *Current Science* **69**(9): 752–763

Lal M, Singh K K, Srinivasan G, Rathore L S, Saseendran A S. 1998. **Vulnerability of rice and wheat yields in NW India to future change in climate**. *Agriculture and Forest Meteorology* **89**: 101–114

Lonergan S. 1998. **Climate warming and India.** In *Measuring the Impact of Climate Change on Indian Agriculture*, edited by A Dinar, R Mendelsohn, R Evenson, J Parikh, A Sanghi, and J McKinsey. Washington, DC: World Bank

MoA (Ministry of Agriculture). 2004. ***Agricultural Statistics at a Glance,*** New Delhi: Directorate of Economics and Statistics, MoA. 126 pp.

MoRD (Ministry of Rural Development). 2001. **Desert Development Programme.** Details available at <http://sirsa.gov.in/htfiles/ddp.pdf>, last accessed on 23 August 2009

Murdiyarso D. 2000. **Adaptation to climatic variability and change: Asian perspectives on agrioculture and food security**. *Environmental Monitoring and Assessment* **61**: 123–131

Oxfam. 2007. **Adapting to climate change: what's needed in poor countries and who should pay.** [Oxfam briefing paper 104. Oxfam International]. Details available at <http://www.oxfam.org.uk/resources/policy/climate_change/downloads/bp104_adapting_to_climate_change.pdf>, last accessed on 13 February 2010

Parry M, Rosenzweig C, Iglesias A, Fischer G, Livermore M. 1999. **Climate change and world food security: a new assessment** *Global Environmental Change* **9**: S51–S67

Rao D G, and Sinha S K. 1994. **Impact of climate change on simulated wheat production in India**. In *Implications of Climate Change for International Agriculture: crop modeling study*, edited by C Rosenzweig, and A Iglesias. Washington, DC: US EPA (United States Environmental Protection Agency)

Rosenzweig C E and Iglesias A. 2006. **Potential impacts of climate change on world food supply: data sets from a major crop modelling study.** [Goddard Institute for Space Studies, Columbia] Details available at <http://sedac.ciesin.columbia.edu/giss_crop_study/>, last accessed on 21 October 2009

Rosenzweig C and Parry M. 1994. **Potential impacts of climate change on world food supply**. *Nature* **367**: 133–38

Rosenzweig C, Parry M L, Fisher G, Frohberg K L. 1993. ***Climate Change and World Food Supply.*** Oxford: University of Oxford, Environmental Change Unit. [Research Report no. 3] Details available at <www.ciesin.org>, last accessed on June 2009

Rosenzweig C, Parry M, and Fischer G. 1995. **World food supply.** In *As Climate Changes: international impacts and implications*, edited by K M Strzepek and J B Smith Cambridge. UK: Cambridge University Press.

Sanghi A, Mendelsohn R, and Dinar A. 1998. **The climate sensitivity of Indian agriculture.** In *Measuring the Impact of Climate Change on Indian Agriculture*, edited by A Dinar, R Mendelsohn, R Everson, J Parika, A Sanghi, K Kumar, J Mckinsey and S Lonergan. Washington, DC: The World Bank [World Bank Technical Paper No 402]

Sinha S K and Swaminathan M S. 1991. **Deforestation climate change and sustainable nutrients security.** *Climate Change* **16**: 33–45

Stern N, Peters S, Bakhshi V, Bowen A, Cameron C, Catovsky S, Crane D, Cruickshank S, Dietz S, Edmonon S, Garbett S L, Hamid L, Hoffman G, Ingram D, Hones B, Patmore N, Radcliffe H, Sathiyarajah R, Stock N, Taylor C, Vernon T, Wanjie H, Zenghelis D. 2006. ***Stern***

Review: the economics of climate change. London: HM Treasury. Details available at <http://www.hmtreasury.gov.uk/independent_reviews/stern_review_economics_climate_change/sternreview_report.cfm>, last accessed on 16 November 2009

Stern N H. 2007. ***The Economics of Climate Change: the Stern review.*** Cambridge: Cambridge University Press.

Venkataraman S and Krishnan A. 1992. ***Crops and Weather. Weather relations of crops.*** pp. 302–457. New Delhi: Indian Council of Agricultural Research

Watson R T, Zinyowera M C, and Moss R H. 2001. **The Regional Impacts of Climate Change: an assessment of vulnerability**. Cambridge, UK: Cambridge University Press. 527 pp.

World Bank. 2006. ***Investment Framework for Clean Energy and Development.*** Washington, DC: The World Bank. pp. 146. Details available at <http://siteresources.worldbank.org/DEVCOMMINT/Documentation/20890696/DC2006-0002(E)-CleanEnergy.pdf>, last accessed on 4 March 2010

World Bank. 2009. ***The Costs to Developing Countries of Adapting to Climate Change. New methods and estimates. The Global Report on the Economics of Adaptation to Climate Change Study.*** Washington, DC: The World Bank. pp. 109 [Consultation Draft]. Details available at <http://siteresources.worldbank.org/INTCC/Resources/EACCReport0928Final.pdf>

five

Coastal zones

G J Lingaraj[a]

Introduction

Coastal regions harbour some of the most diverse and productive ecosystems and act as an active interface between land and water. India has a long low-lying coastline extending over 7500 km across nine states, covering about 70 coastal districts (Census of India 2001).[1] The average population density in the coastal districts is 455, which is about 1.5 times the national average. Three metropolises—Mumbai, Chennai, and Kolkata—and 18 cities with more than a million population are located in the coastal zones. There are 11 major and 130 minor seaports,[2] and about 1551 industries are located along the coastline.[3] The presence of sandy beaches, coral islands, backwaters, and the United Nations Education, Scientific, Cultural Organisation (UNESCO) world heritage sites have led to strong tourism activity all along the coastline. Most of the coastal regions are agriculturally fertile, with many rice-growing belts in the states of Andhra Pradesh, Kerala, Orissa, Tamil Nadu, among others. Roughly, about

[a] Research Associate, Centre for Global Environment Research, Earth Science and Climate Change Division, The Energy and Resources Institute (TERI), New Delhi, India.

[1] Details available at <http://censusindia.gov.in/>, last accessed on 9 December 2009.

[2] Details available at <http://shipping.gov.in>.

[3] GoI (1997) reporting to the UN Commission on Sustainable Development. Details available at <http://www.un.org/esa/earthsummit/india-cp.htm#chap17>.

100000 hectares (ha) of land is being used for prawn culture in the eastern coast. Sizeable area of coastal land, particularly, within the reach of high tide, and also mangrove areas are being used for aquaculture.[4]

Different parts of the coastline have different characteristics. The eastern coastline is neutral and highly dynamic along the deltas of Ganga, Mahanadi, Krishna, Godavari, and Cauvery rivers due to high sedimentation, whereas the western coast is a submerging coastline interspersed with coastal rivers, inland creeks, and backwaters. The eastern coast is rich in mangrove forests. Mangrove forests are concentrated in the Sundarbans, Mahanadi delta, Coringa, and Andaman and Nicobar Islands. Coral reefs are predominantly found in the islands of Lakshadweep and Andaman and Nicobar and also in the small islands in the Gulf of Kachchh and Gulf of Mannar. Coastal ecosystems are not only a rich source of biodiversity, but they also play a crucial role in protecting the coastal zones from erosion.[5]

Coastal regions are highly productive systems. The development activities in the coastal zones have not only mounted huge stresses on the existing resource base (Mani 2000; ADPC 2005; Jagtap, Kulkarni, and Verlekar 2008) but have also worsened some of the existing problems like degradation of coastal ecosystems, erosion, and flooding.

Coastal regions are highly sensitive to climate-related stimuli owing to their exposure to sea level rise (SLR) and tropical cyclones. The *Fourth Assessment Report (AR4)* of the Intergovernmental Panel on Climate Change (IPCC) predicts an increased SLR of between 0.18 m and 0.59 m across various emission scenarios over the next 100 years (IPCC 2007). It also projects, with a high degree of confidence, an increase in the intensity and frequency of extreme weather events, including storm surges and cyclonic events. The projected biophysical changes due to climate change could have severe adverse socio-economic implications for human settlements,

---

[4] GoI (1997) reporting to the UN Commission on Sustainable Development. Details available at <http://www.un.org/esa/earthsummit/india-cp.htm#chap17>.

[5] For an overview of the extent of coastal ecosystems in India, refer Annexure I.

health, tourism, freshwater resources, agriculture, fisheries, and so on (IPCC 2007; McLean, Tsyban, Burkett, *et al.* 2001; GoI 2004; Nicholls and Small 2002). High climate sensitivity and low adaptive capacity of the coastal regions in India make the coastal zone a highly vulnerable system.

In this regard, it is imperative to carefully plan and devise comprehensive adaptation strategies to combat the adverse impacts of climate change. Adaptation will incur costs and will not mitigate all damages. IPCC (2007) observes, with a high degree of confidence, that adaptation in the coasts of developing countries will be much more challenging than that in the coasts of developed countries. It also adds, with high degree of confidence, that costs of adaptation for vulnerable coasts are much lesser than the costs of inaction. Therefore, estimates of adaptation costs are essential for taking informed policy decisions on adaptation.

This study is an attempt to estimate the costs of adaptation to climate change in the coastal regions of India. It also attempts to arrive at a methodological framework based on a literature review of the following.

- Impacts of climate change on the coastal regions with a focus on India.
- Adaptation responses to the impacts of climate change, along with suitable examples of relevant policies and programmes from India.
- Methodological approaches to assess costs of adaptation in the coastal regions.

Based on the methodology developed, we assessed the costs of adapting to climate change in the coastal regions of India. We conclude the chapter by analysing the findings, identifying gaps in current knowledge, and enumerating the limitations of the study.

Impacts of climate change on the coastal zones in India

Most coastal regions will be subject to increased levels of flooding, leading to loss of land due to inundation, accelerated erosion, salt-water intrusion, and loss of wetlands and mangrove ecosystems. The key climate-related risks associated are SLR, flooding, storm surges, and tropical cyclone occurrences as well as changes in precipitation and sea-surface temperatures (IPCC 2007). While most of the literature has focused on studying the impacts of SLR on the coastal regions (Nicholls and Small 2002) and its socio-economic effects (Mendelsohn, Morrison, Schlesinger, *et al.* 2000; Ng and Mendelsohn 2005), there are very few studies that have taken into account other climatic risks and their consequent impacts. Table 1 illustrates the significance of other climatic factors.

Table 1 Bio-geophysical effects of climate change on the coastal zones

Climate factor	Direction of change	Bio-geophysical effects
Sea water temperature (of surface waters)	Increase	Increased coral bleaching; migration of coastal species towards higher latitudes; decreased incidence of sea ice at higher latitudes
Precipitation intensity/ run-off	Intensified hydrological cycle, with wide regional variations	Changed fluvial sediment supply; changed flood risk in coastal lowlands; catchment management
Wave climate	Poorly known, but significant temporal and spatial variability expected	Changed patterns of erosion and accretion; changed storm impacts
Storm track, frequency, and intensity	Poorly known, but significant temporal and spatial variability expected	Changed occurrence of storm flooding and storm damage
Atmospheric CO_2	Increase	Increased productivity in coastal ecosystems; decreased $CaCO_3$ saturation impacts on coral reefs

Source Nicholls and Small (2002)

Effects like coral bleaching and mangrove degradation will increase the exposure, and increase in the frequency and intensity of storms will increase the sensitivity, leading to an overall increase in vulnerability of the coastal zones. This section reviews the literature on the impacts of climate change on various socio-economic sectors in the coastal zones of India. Figure 1 schematically represents the various impacts and the pathways through which climate change would impact coastal zones.

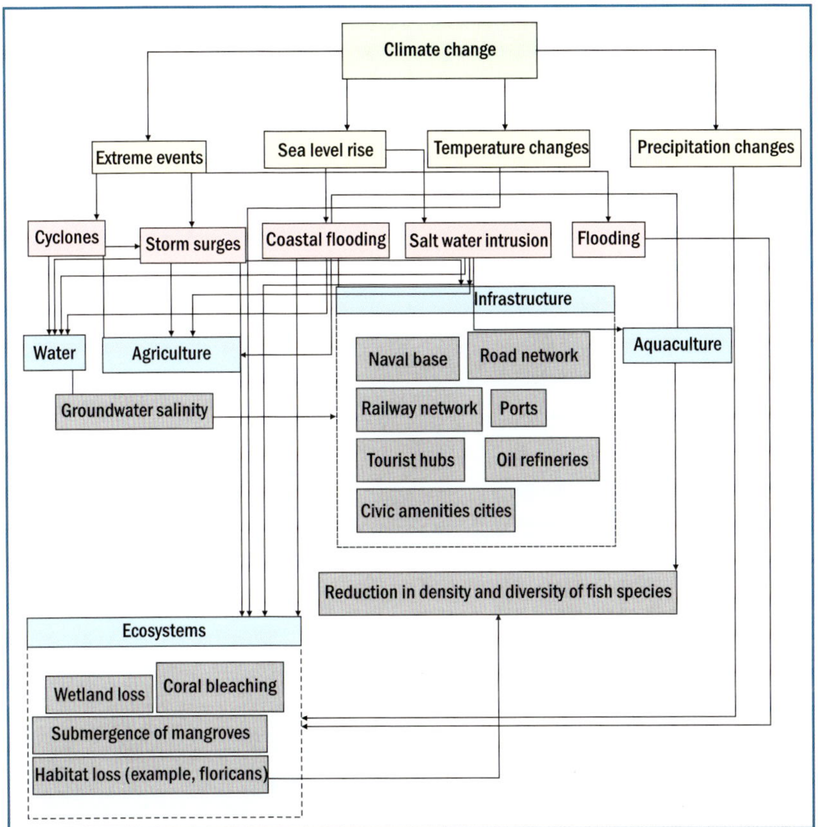

Figure 1 Climate change impacts on the coastal regions of India

Impacts of sea level rise

A variety of impacts are expected as a consequence of SLR. The primary effects include increased coastal flooding (Hoozemans, Marchand, and Pennekamp 1993), loss of land due to inundation (Fankhauser 1994; Nicholls 2000), and erosion of sandy beaches (IPCC 1992). These could lead to loss and degradation of ecosystems such as wetlands, mangroves (Field 1995), and corals (Nicholls and Small 2002); degradation of coastal vegetation and habitats; and intrusion of salt water into groundwater systems. Inundation due to SLR could have huge implications on human systems and settlements (Nicholls 2000), leading to the displacement of population (Yohe 1990), loss of life and economic losses (Nicholls and Tol 2006; Mendelsohn, Morrison, Schlesinger, *et al.* 2000; Asthana 1994; Norahna, Choudri, and Nairy 2003).

SLR leads to the intrusion of saline water into the fresh groundwater in coastal aquifers, adversely affecting the groundwater resources. Impacts of SLR are also felt prominently through changes in the other sectors, such as changes in food production and water availability (Mitra 2004). Large-scale migration due to submergence of coastline as a result of SLR can be expected (Yohe 1990; Alam 2003). With its low-lying, densely populated coastline, India is extremely vulnerable to SLR. Erratapuzha and George (1980) observed through remote sensing and geographic information system (GIS) that there has been a considerable recession in the Kerala coast between 1860 and 1966. Shetye, Gouveia, and Pathak (1990) observed that coral islands of Lakshadweep are highly vulnerable to SLR. A 1-m rise in sea level is expected to inundate about 1700 km^2 of agricultural land in Orissa and West Bengal (IPCC 1992).

There are some global (IPCC 1992; Fankhauser 1994; Tol 1995, 2002; Nicholls and Tol 2006; Toman 2006) and Indian studies (ADB 1994; TERI 1996) on the economic costs of SLR. The Asian Development Bank (ADB) conducted a vulnerability assessment of the 1-m SLR for the India's coastal zones using a GIS analysis tool based on topographic maps and socio-economic data (ADB 1994). It estimated that the economic cost due to land loss, population displacement, and impact on the forestry was \$71 292 million if it occurred in 1994. TERI (1996) estimated that in the absence of protection, a 1-m SLR would affect a total area 1563 km^2 and put 7.1 million people at risk. The dominant cost would then be due to land loss of over 5764 km^2. SLR will also adversely affect the tourism sector. It is likely to suffer a loss of Rs 19 635 000 million by 2050 (\$419 million at an exchange rate of Rs 46.86 to the dollar).

Agricultural impacts

Agricultural production in several coastal areas is heavily dependent on climatic factors. Exposure to extreme climatic shocks like storms and severe storms results in a negative effect on agriculture and the incomes of people. The net impacts of climate change on agricultural output in the coastal zones of India are uncertain, yet specific regions and certain groups of farmers, particularly those farming on marginal, rain-fed lands, are likely to suffer significant damages as a result of climate change. Most of India's coastal regions are fertile and under paddy cultivation, which is sensitive to inundation and salinization (GoI 1981). The eastern coastal districts are major producers of rice in India. West Coast agricultural regions, including Gujarat, Maharashtra, and Karnataka, are expected to be among the most negatively affected by climate change. Coastal drought-prone areas include Kachchh, central coastal Andhra Pradesh, southern Tamil Nadu, Saurashtra, and south-western West Bengal. Climate change could worsen the situation in these drought-affected states. The cyclones that struck several states of India in 1952, 1955, 1996, and 1998 destroyed coconut palms, and it will take years to restore their production to the pre-cyclone levels (GoI 2004).

Impacts on ecosystems

Salinity intrusion, SLR, and changes in sea-surface temperature and pH have significant impacts on the coastal ecosystems (Nicholls and Small 2002). SLR would submerge beaches, salt marshes (ENVIS Centre 2007), mangrove forests, and mudflats, leading to habitat loss of many species such as the Olive Ridley turtle nesting sites on the East Coast, Flamingos and lesser floricans in the Rann of Kachchh, and the Bengal Tiger in the Sundarbans (Smith, Tol, Ragland, et al. 1998). These are all severely threatened by SLR. Submergence of mangroves might also lead to the degeneration of salt-intolerant mangrove species. It has been observed in Satavaya, Orissa, that the sea has ingressed to about 1.5 km, leading to a loss of 56% of its mangrove vegetation. The rising sea levels would have serious implications for the Sundarbans[6] region, which is a habitat for a rich diversity of mangroves. A study by Smith, Tol, Ragland, *et al.* (1998) estimated that a 45-cm SLR would inundate 75% of the Sundarbans, and a 65-cm SLR would inundate the entire mangrove ecosystem there. Mangrove degradation increases vulnerability to cyclones (MoEF 2007). Mangroves are also impacted by changes in local temperature and precipitation, which influence the salinity with a bearing on the plant composition.

Saline water intrusion, coupled with changes in tidal pattern, will adversely affect 260 species of fish in the Sundarbans region, which

---

[6] Sundarbans is the largest delta (delta of Ganga and Brahmaputra) in the world and is shared between two neighbouring countries of Bangladesh and India. It also contains the largest mangrove forest in the world.

are sensitive to particular salt and freshwater conditions. Over the past two decades, there has been a rapid increase in the number of bleaching events. In Gulf of Mannar, Tamil Nadu, it was reported that coral bleaching has increased from 14.9% to 15.5% from 2005 to 2006. The corals of the Lakshadweep are also significantly affected, with bleaching of over 80% of coral cover and 25% coral mortality.

Impacts of cyclones

Tropical cyclones have major economic, social, and ecological consequences for the coastal areas (Shrestha 1998; IPCC 2007). On an average, about 119 million people in India are exposed to a tropical cyclone hazard every year. According to the database compiled by the Centre for Research on the Epidemiology of Disasters (CRED),[7] there were about 138 cyclonic events, affecting about 87 516 000 people and causing damages of about $12 707 million during 1900–2005. Based on the historical analysis of the spatial pattern of cyclone incidence during 1877–1990, Patwardhan, Narayanan, Parthasarathy, *et al.* 2003 observed that on average about 8.45 cyclones cross the Indian coastline every year. The eastern coast is more vulnerable with respect to the frequency and intensity of cyclones (Patwardhan, Narayanan, Parthasarathy, *et al.* 2003; Patwardhan and Narayanan 2003). The 1999 super cyclone of Orissa resulted in massive destruction to life, property, and infrastructure. More than 10 000 lives were lost, and over a million were rendered homeless (CRED database).

IPCC (2007) and GoI (2004) observe that there would be an increase in the intensity and frequency of cyclones in the future. Lal (2001) suggests that cyclone intensity could possibly increase by 10%–20% for 2–4°C rise in the sea surface temperature. Based on regional HadCM2[8] simulations, Unnikrishnan, Rupa Kumar,

---

[7] The Centre for Research on the Epidemiology of Disasters (CRED) is a research unit of the Université catholique de Louvain (UCL). It is situated at the School of Public Health of the Brussels campus of the university. CRED has been active for over 30 years in the fields of international disaster and conflict health studies, with research and training activities linking relief, rehabilitation, and development.

[8] HadCM2 is a coupled global model developed by Hadley Centre, England. Details available at <http://cera-www.dkrz.de/IPCC_DDC/IS92a/Hadley-Centre/hadcm2.html>.

Fernandes, *et al.* (2006) predict increases in the frequency as well as intensities of tropical cyclones by 2050 under the IS92a[9] scenario in the Bay of Bengal.

Adaptation measures in the coastal zones

Adaptation is defined by the IPCC as the "adjustment in natural or human systems in response to actual or expected climatic stimuli or their effects, which moderates harm or exploits beneficial opportunities" (Adger, Agrawala, Mirza, *et al.* 2007). They can be classified as proactive and reactive or anticipatory and planned, amongst other depending on the criteria chosen (EEA 2007). Bijlsma, Ehler, Klein, *et al.* (1996) identified three broad types of adaptation strategies for the coastal zones. These are as follows.

1 *Protect* This is essentially a structural intervention involving protection of coastal areas with hard and soft barriers. It involves putting up of artificial physical structures in the landscape. Hard structural measures include construction of dykes; soft measures include mangrove regeneration, shelterbelt plantations, and wetland conservation. Coastal landforms (for example, coral reefs and barrier islands) and wetland ecosystems (for example, marshes and mangroves) provide a natural first line of protection from storm surges and flooding despite divergent views about the extent to which they reduce destruction (Nicholls, Wong, Burkett, *et al.* 2007). Mangroves, for example, protect coastal areas against strong winds and cyclones, and also check soil erosion. The importance of coastal plantations and mangrove forests in mitigating the impact of climatic uncertainties like cyclones has been well documented (Das and Vincent 2009; Massel 1999).

2 *Accommodate* It permits continued occupancy and use of vulnerable areas with non-structural interventions. It involves land-use controls that allow for coastal setbacks, better information dissemination, and economic incentives to reduce

---

[9] IS92a is a middle of the range scenario in which population rises to 11.3 billion by 2100, economic growth averages 2.3% a year between 1990 and 2100, and a mix of conventional and renewable energy sources are used. Only those emissions control internationally agreed upon and national policies enacted into law.

or prevent disasters. Examples of this in the Indian context could be the use of coastal regulation zones (CRZs) and implementation of community-based disaster risk management initiatives like the Disaster Risk Management[10] Programme of the Government of India (GoI)–United Nations Development Programme (UNDP), which envisages capacity development for disaster reduction at the national level and in some of the most vulnerable regions in the country through community-based and gender-sensitive approaches. The programme has enhanced awareness and built capacity amongst the masses.

3 *Retreat* It involves temporary or permanent abandonment of land and structures in vulnerable areas and resettlement of inhabitants. This involves preventing development in the vulnerable area or allowing development to take place on the condition that it will be abandoned, if necessary, on a planned phase out. A good example for such an adaptive measure in India is the 1991 CRZ Notification, wherein no new construction is permitted in zone 1, which is within 500 m of the high tide line.

Some of the important policies and programmes in India corresponding to adaptation options are mapped in Table 2 with the relevant impacts.

Table 2 Mapping of impacts of climate change on the coastal zones with adaptation options and baseline adaptive responses

Impacts		Adaptation options	Policy/programme
Inundation flood/ storm damage	Surge (sea)	Dikes/surge barriers (P)	Shelter belt plantation
			Construction of embankments
		Building codes/ flood wise buildings (A)	Coastal regulation zone (CRZ) notification construction guidelines in coastal regulation zone III

Continued...

[10] Based on good practices in community-based disaster risk management, Disaster Risk Management Programme of the Government of India (GoI)–United Nations Development Programme (UNDP).

Table 2 *Contd...*

Impacts		Adaptation options	Policy/programme
	Backwater effect (river)	Land use planning/ hazard delineation (A/R)	CRZ Notification, National Disaster Management Act, 2005
Wetland loss (change)		Land use planning (A/R)	CRZ Notification
		Managed realignment/forbid hard defences	
		Nourishment/ sediment management (P)	
Erosion (direct and indirect change)		Coast defences (P)	Shelter belt plantation
		Nourishment (P)	Mangrove regeneration as a part of mangroves for future programme
		Building setbacks (R)	
Saltwater intrusion	Surface	Saltwater intrusion barriers (P)	CRZ notification
		Change water abstraction (A)	
		Freshwater injection (P)	
	Groundwater	Change water abstraction (A)	
Cyclones		Cyclone shelters (A)	National Cyclone Risk Mitigation project, 2004
		Disaster planning and efficient early warning systems (A)	National Disaster Management Act, 2005
Ecosystem degradation	Corals	None	None
	Mangroves	Mangrove regeneration	Integrated Afforestation and Eco-development Project Scheme (2001)[a]

Note Alphabets in the brackets denote the type of adaptation, with P, A, and R referring to protect, accommodate, and retreat, respectively.
A – accommodate; P – protect; R – retreat; CRZ – coastal regulation zone
[a]Details available at <http://envfor.nic.in/report/0001/chap06.html#rd_naebiaep>, last accessed on 3 August 2009.

All these adaptation measures should be intertwined with integrated coastal zone management[11] strategies and should be consistent with national, state, community, and private goals.[12]

Costing adaptation: methodological issues

Coastal regions have received relatively more attention than the other sectors owing to their greater vulnerability to climate change through SLR. Quite a number of studies have tried to quantify the impacts of SLR (Frankhauser 1994; Hoozemans, Marchand, and Pennekamp 1993; Leatherman and Nicholls 1995; Nicholls 1995; Nicholls and Hoozemans 1996; Yohe, Neumann, and Ameden 1995; Yohe and Schlesinger 1998; Nicholls and Mimura 1998), and adaptations cost estimates in the coastal zones are available from early 1990s.[13] The first few studies on estimation of the cost of SLR looked at response strategies for holding back the sea and not holding back the sea (Yohe 1990; Titus, Park, Leatherman, *et al.* 1992; Bijlsma, Ehler, Klein, *et al.* 1996)

Owing to methodological challenges associated with data availability, cross-sector linkages, non-linear impacts, and feedbacks, amongst others, almost all adaptation cost estimation methods follow a similar simplistic approach. On the impacts side, they consider the impacts of SLR only, and on the adaptation side, the focus is on coastal protection. The cost of SLR is usually divided into three categories: capital costs of protective infrastructure, costs associated with dry land loss, and costs of wetland loss. This can be given as shown below.

Cost of sea level rise = costs of dry land loss + costs of wetland loss + protection costs

---

[11] This concept was born in 1992 during the Earth Summit in Rio de Janeiro. It involves an integrated approach to manage the coasts, giving due recognition to all aspects of the coastal zone, including geographical and political boundaries, in an attempt to achieve sustainability.

[12] For an overview of policies and programmes in India relevant to coastal regions, refer Annexure II.

[13] Agrawala and Frankhauser (2008) provide a critical overview of adaptation costing methodology.

The models have differed considerably in deciding what is the optimal level of protection; in terms of what to protect and how much to protect? The global vulnerability analysis of Hoozemans, Marchand, and Pennekamp (1993), one of the first studies carried out on adaptation costs in coastal zones, considered all low-lying areas for protection, and the level of protection was fixed on the basis of per capita income. Countries with per capita income over $2400 were to be protected against 1000-year storm, while countries with incomes per capita of $2400–$600 were assumed to require protection against 100-year storm, and countries with per capita incomes below $600 were to be protected against 10-year storm. Nicholls and Small (2002) and Nicholls and Lowe (2004) followed the same approach but altered the thresholds. Fankhauser (1994) and Yohe, Neumann, and Ameden (1995) used cost–benefit analysis to find out the optimal levels of protection under which the optimal level of coastal protection is attained when the marginal cost of protection from SLR is equal to the marginal benefit of protection. However, there is one crucial difference between the approaches of Fankhauser and Yohe. While Fankhauser considers the entire coastline of a country, Yohe applies cost–benefit analysis for selected fragments of the coast line.

Methodology

In order to delineate our scope, it is important to define coastal zones. A coastal zone, in its simplest form, could be defined as that part of the land, which is affected by its proximity to the sea, and that part of the ocean, which is affected by its proximity to the land (US Commission on Marine Science, Engineering, and Resources 1969). The Swaminathan Committee[14] (2007) defines a coastal zone as an area with territorial limits (12 nautical miles), including its sea-bed to the administrative boundaries or the biological boundaries demarcated on the landward side of the sea coast and would also include inland tidal waterbodies influenced by tidal action and the land area along such waterbodies. These definitions highlight the

[14] Swaminathan Committee Report was set up to review the working of coastal regulation zone in India. Details available at <http://envfor.nic.in/news/janmar05/swaminathan.pdf>, last accessed on 4 August 2009.

dynamic nature of the coasts, influenced both by the processes of land and sea. In the present study, the district was chosen as the unit of analysis, due to data availability at that level. A detailed socio-economic, environmental, and hazard profiling of all the coastal districts was carried out based on secondary studies and historical data. The profiling involved identifying wetlands, mangrove hubs, ports, industrial, and tourism belts. Based on historical records of frequency and intensity of cyclones, districts were classified into high-, medium- and low-risk categories. Flood-prone districts were identified based on historical records. The objective of profiling was to identify the relevant adaptation strategy and its spatial distribution. Figure 2 provides the main characteristics of the coastal districts in India.

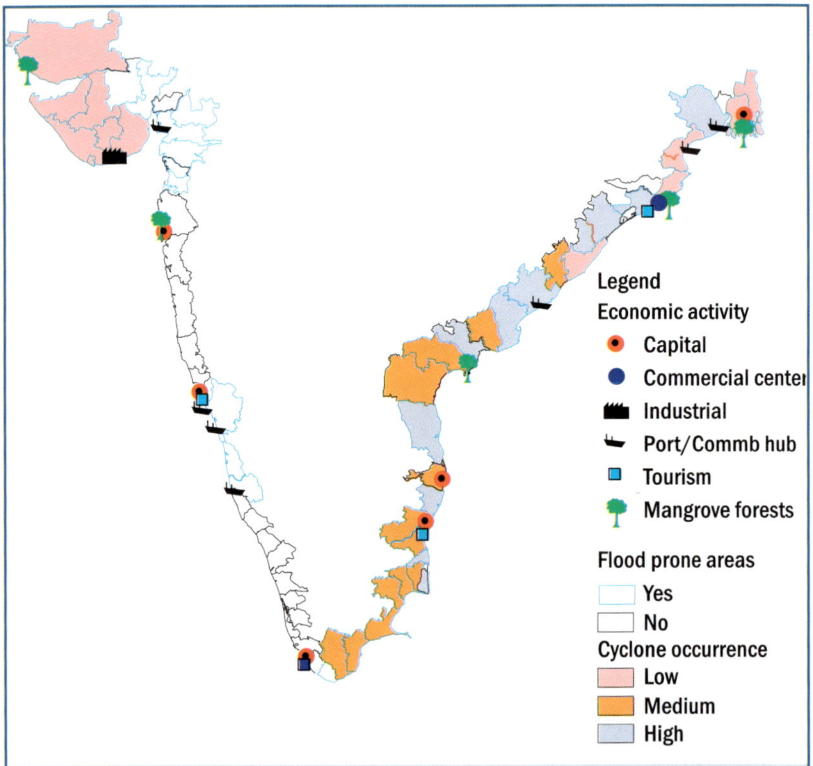

Figure 2 Ecological, socio-economic, and hazard profiling of coastal districts

Scope

The time frame considered for the study is 2030, and the scenario chosen is A2.[15] For this scenario, IPCC (2007) projects an accelerated SLR of about 0.3 m for the 2030s. Given the time frame of the study and the projected impacts for that period, a mix of retreat, accommodation, and protection, the three broad types of adaptation strategies discussed in the earlier section, would be required. For instance, temporary or seasonal retreat during intense storms, accommodation through better coastal zone regulations, disaster warning, insurance, and so on, as well as coastal protection could be employed. The study limits itself to planned adaptation and does not consider autonomous adaptation because the impacts of climate change on the coastal zones are primarily beyond the coping capacity of the individuals and would require planned interventions. Adaptation options considered in the study are as follows.

- Coastal protection
- Beach nourishment[16]
- Shelterbelt plantation, which also includes mangrove regeneration
- Cyclone risk mitigation through cyclone shelters

Approach

Although the study follows the conventional approach in assessing the optimal level of coastal protection due to SLR and subsequent costs associated with that, it also makes an attempt to overcome the limitations of that approach by including cyclone risk mitigation in the analysis.

The starting point of most of the previously conducted studies is an SLR scenario (usually of 1 m) on the basis of which the cost of protecting the coasts is calculated. The study builds upon the earlier

---

[15] The A2 scenario is characterized by a world of independently operating, self-reliant nations, continuously increasing population, regionally oriented economic development, and has lower and more fragmented technological changes and improvements with respect to per capita income.

[16] Beach nourishment is a process by which sediment (usually sand) lost through long shore drift or erosion is replaced on a beach. It involves the transport of the nourishment material from one area to the affected area.

methodologies and intends to complement the top-down approach currently followed to calculate the costs for coastal protection with a bottom up unit cost approach that takes into account other adaptation measures as well.

The data used in the Dynamic Interactive Vulnerability Assessment (DIVA) model is given in McFadden, Vafeidis, and Nicholls (2003). The GoI figures have been used at most of the places. We describe the methodology employed to assess costs for each of the components in the subsequent sections.

Coastal protection and beach nourishment

The DIVA tool was used to calculate the costs for coastal protection. DIVA is an outcome of a European Commission (EC) funded multi-institutional project called the Dynamic and Interactive Assessment of National, Regional and Global Vulnerability of Coastal Zones to Climate Change and Sea-Level Rise (DINAS-COAST).[17] One of the key outcomes of this project is a software tool,[18] which is used to assess coastal vulnerability and generate quantitative information on a range of coastal indicators. The tool comprises a global database of biophysical and socio-economic data on the coastal zones. It also comprises global and regional climate and socio-economic scenarios up to 2500. Annexure III contains details of physical features of coastal zones in India, considered in DIVA. The schematic given in Figure 3 represents the structure and working of the DIVA model.

A set of interacting algorithms containing modules on the impacts of and adaptation to SLR, coastal erosion, and flooding computes and analyses 12 418 coastal segments that constitute coasts across the world (Nicholls, Wong, Burkett, *et al.* 2007; Vafeidis, Nicholls, McFadden, *et al.* 2008). Coastal segments are parts of coastline sliced based on select indicators. The model not only considers direct effects of SLR but its indirect effects on coastal wetlands as well. SLR scenarios are downscaled for each coastal segment. The impacts of SLR are analysed using Bruun Rule (1962) (Box 1) (Bruun 1962).

---

[17] Details available at <http://www.dinas-coast.net/>.

[18] DINAS-COAST Consortium (2004). (DIVA 1.0. Potsdam Institute for Climate Impact Research, Potsdam, Germany, CD-ROM).

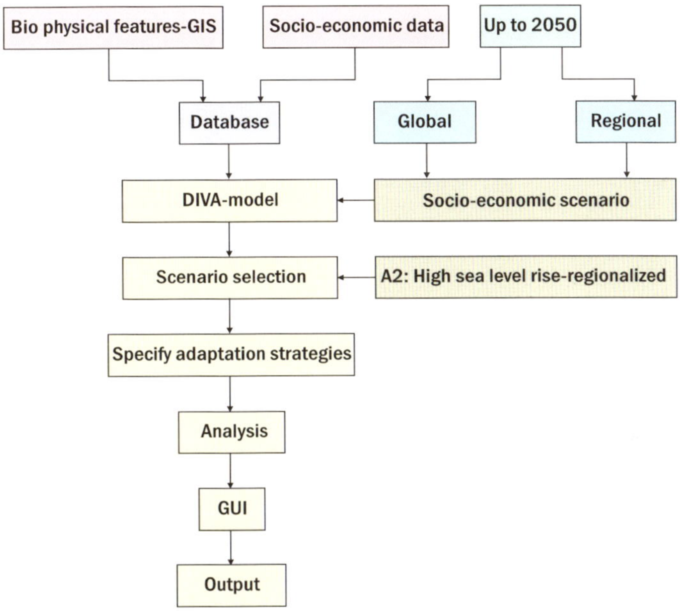

Figure 3 Working of the Dynamic Interactive Vulnerability Assessment (DIVA) model

Box 1 What is Bruun rule?

Bruun rule is a shoreline response model to accelerated sea level rise (SLR), originally proposed in 1954 by P Bruun. The rule states that a marine shoreline maintains an equilibrium profile with a depth and slope, which are determined by the current and wave regime. It is a simple two-dimensional model of shoreline response to accelerated SLR. Bruun rule is applicable only to sandy shores and straight areas of the coastline.

To maintain the profile shape, the upper part of the profile erodes and the lower part of the profile accretes, translating the shoreline landward. The shoreline recession due to SLR can be calculated as shown in Equation 1.

$$R = G\,(L/H)\,S \qquad \ldots (1)$$

where R is the shoreline recession; S is the shoreline, H is the active profile height; L is the active profile width between boundaries; and G represents the grain size of eroded material, that is, material that is too fine for the beach is lost, increasing the predicted erosion.

Continued...

Box 1 *Contd...*

The active profile height can be calculated as shown in Equation 2.

$$H = B + h^* \quad \text{... (2)}$$

where h^* is the depth at the offshore boundary, which is termed the depth of closure and B is the appropriate land elevation.

The inverse of the beach slope (L/H) multiplies any SLR, giving a large recession for a small rise in sea level. Therefore, flatter beaches are predicted to show greater sensitivity to SLR than steeper beaches.

Bruun rule has been subjected to a lot of criticisms arising out of simplistic assumptions, such as the assumption that the shore is composed only of sandy beaches and loses no sediment. In reality, shorelines are composed of rocks and beaches amongst other things. Another significant drawback of this rule is that it cannot be applied to assess shoreline responses in short time scales (<10 years). Another significant demerit of this model is that it ignores significant site-specific variables. However, recent work by Stive (2004) and Zhang, Douglas, and Leatherman (2004) shows that Bruun rule is by far the best available approach to understand coastline response to accelerated SLR.

The model estimates damages due to the impacts of SLR and the consequent costs for adaptation. Physical impacts and adaptation options considered in the model, along with cost estimation sources, are listed in Table 3.

Table 3 Impacts and adaptation options considered in the model

Physical impact of SLR	Adaptation approach	Cost estimation source
Inundation, flood, and storm damage	Hard protection (via dykes)	Global Vulnerability Analysis (Hoozemans, Marchand, and Pennekamp 1993)
Wetland loss (and change)	Sediment nourishment	
Erosion (direct and indirect morphological change)	Beach nourishment	Environment Protection Agency
Saltwater intrusion	Adaptation not considered	

SLR – sea-level rise

In the DIVA model, a cost–benefit analysis for each coastal segment is employed to assess optimal levels of coastal protection. Both bio-physical and socio-economic damages are assessed based on select indicators. Table 4 lists the adaptation options considered in the study and the basis on which they were included.

Table 4 Details of adaptation options considered in the model

Option	Details
Protection mode for sea floods	Minimum 100 years
Beach nourishment mode	Cost-benefit analysis
Tidal basin nourishment mode	Cost-benefit analysis
Wetland nourishment mode	Cost-benefit analysis

Infrastructure is not explicitly defined in the model's database, but is estimated via population distribution and gross domestic product (GDP). The flooding model assesses temporary inundation of land areas and associated values, while erosion deals with damage to tourism and infrastructure.

DIVA implements the adaptation options according to complementary adaptation strategies. For beach nourishment, a cost–benefit adaptation strategy balances costs and benefits (damage avoided) of adaptation, including the tourist value of beaches. For a dike building, the demand function for safety is applied over time, subject to population density. The demand for safety function increases with per capita income and population density and decreases with the costs of dike building, an approach that is posited as the solution to a cost–benefit analysis (Tol, Bohn, Downing, *et al.* 2006). Dikes are built only when population density exceeds 1 person/km^2, with an increasing proportion of the recommended height being built as population density rises; for example, 98% of the dike height is built at densities of 1000 people/km^2.

The model enables the interaction between modules that assess biophysical and socio-economic impacts and the potential effects and costs of adaptation for the user-defined scenarios. SLR scenarios are downscaled with an estimate of the vertical land movement in each segment. The coastal erosion analysis considers only sandy coasts and takes into account the direct effect (Bruun effect) and indirect effects of SLR, as well as beach nourishment, wherever relevant.

The indirect effects occur at major estuaries and lagoons. The flooding analysis determines the flood areas for different return periods and extreme water levels, including the effects of dikes.

This tool includes flood and erosion models, which estimate both damage and costs in response to planned adaptation in the form of dike construction and beach/shore nourishment.

Shelterbelt plantations

A large proportion of the coastal population in India is dependent on coastal and marine resources like mangroves for their livelihoods directly or indirectly. Coastal ecosystems such as mangroves also act as natural shield for the coasts by diffusing the energy and intensity of the waves reaching the shore (Badola and Hussain 2005; Balaji 2005). The majority of the mangroves in India are found along the East Coast, where rivers and estuaries are more extensive, than on the West Coast. The Gangetic delta mangroves, the Sundarbans, formed at the confluence of the Ganga and the Brahmaputra are the world's most extensive continuous mangrove forests and cover about 2000 km^2 of the Indian territory (Blasco and Aizpuru 2002).

Mangroves protect coastal areas against strong winds and cyclones, and also check soil erosion (Mazda, Magi, Kogo, *et al.* 1997; Massel 1999). The tsunami of 2004 clearly demonstrated the importance of mangrove forests and coastal plantations. These ecosystems also serve as a rich repository of biodiversity and provide waste assimilation services. In India, however, they are amongst the most threatened natural ecosystems.

Although the mangrove conservation programme was launched way back in 1987, the Planning Commission Task Force on Islands, Coral Reefs, Mangroves, and Wetlands in 2007 noted that there has been a significant degradation of mangrove forests over the last few decades. Das (2007) noted in her study on the Orissa Super Cyclone of 1999 in Kendrapada,[19] that there would have been a 92% reduction in damages, had the mangroves been at 1950 levels. Coastal ecosystem degradation not only has increased the exposure of the coastal zones to storm surges, cyclones, and tsunamis but also

---

[19] Kendrapada is a district in coastal Orissa.

has a direct adverse impact on coastal livelihoods, which further increases their vulnerability (Mani 2000).

To increase the resilience of the coastal communities to the above-mentioned hazards, the Ministry of Environment and Forests (MoEF) has been encouraging mangrove regeneration and shelterbelt plantations under Integrated Afforestation and Eco-development Project, and other programmes (MoEF 2002). Shelterbelt plantations are bioshields formed by planting a vegetation belt along the coastal zones to protect against erosion, cyclones, and tsunami (Ramesh and Ramachandran 2007). These shelterbelts are composed of mangrove species or tree species like Casuarina (Selvam, Ravishankar, Karunagaran, *et al.* 2005), protection capability of which was evident during the Indian Ocean tsunami of 2006 (Balaji 2005). Along with storm protection, these shelterbelts also help in nutrient retention, fish and shellfish production, and erosion prevention. The coastal shelterbelt plantation and mangrove rehabilitation are also a part of other programmes like National Afforestation Programme Scheme (1997), National Cyclone Risk Mitigation Project (2004), and Tsunami Rehabilitation Programme (2008).

Per hectare costs of mangrove regeneration and casuarina plantation are similar only under certain conditions (Selvam, Ravishankar, Karunagaran, *et al.* 2005). Even with same mangrove species, the costs are not same at different locations due to different physiographic conditions like climate, salinity, tidal flows, wind velocity, and soil, which are different in different regions. Since afforestation is a labour-intensive task, costs also vary with labour costs. Given the data limitations, we have been obliged to use median values of costs.

It is difficult to establish the optimum spread and composition of these afforestation drives. This is primarily due to two reasons. First, mangrove species have a very narrow ecological threshold unlike casuarina, which thrives almost everywhere. Since the costs per hectare of the two were the same, it was assumed that regions conducive for mangroves will have mangrove regeneration, and the rest of the regions will have casuarina plantations. Second, to meet this methodological challenge of the optimal spread of afforestation drive, the study identified the total area of wastelands in the coastal districts as designated by the Planning Commission of the GoI[20] and selected it as the area where afforestation could be carried out. The total cost of shelterbelt plantation was the product of per hectare cost of afforestation and total wasteland area in the coastal zones.[21]

Cyclone risk mitigation

Cyclones pose an immediate and significant threat to the coastal zones. Economic damage of cyclones in India during 1970–2000 is estimated at $7.8 billion (CRED Database). As noted earlier, the frequency and intensity of tropical cyclones are projected to increase in future. Therefore, any estimate of adaptation costs in the Indian context is incomplete without including cyclone risk mitigation in the analysis. This is a significant drawback of the DIVA model, which limits itself to beach nourishment and coastal protection in its analysis. It is not an easy task to consider cyclone risk mitigation in the analysis because of the methodological challenges it poses. Although we have studies on the projected increase in the intensity and frequency of cyclones (Henderson–Sellers, Zhang, Berz, *et al.* 1998; GoI 2004), there is a fair bit of uncertainty on the timing and the magnitude of this increase. It, therefore, becomes difficult to assess adaptation costs, since most of the assessment methods are impact-centric. There are severe data limitations as well, which further constrained the scope of this exercise. Nevertheless, we have attempted to make an estimate of the costs, based on an approach that is described subsequently.

Cyclone risk mitigation through cyclone shelters

It has been noted that in the near future, reducing vulnerabilities is the best way to mitigate the effects of cyclones (Pielke 2007). Construction of cyclone shelters has been a proven means of preparedness (NDMA 2008) as the vulnerable populations can be evacuated to these structures immediately after receiving the cyclone warning. The Orissa State Disaster Management Authority

---

[20] Details available at <http://planningcommission.nic.in/plans/planrel/fiveyr/10th/volume2/v2_ch5_3.pdf>, last accessed on 21 January 2010.

[21] Please refer Annexure IV for details of the calculation.

(OSDMA), in its assessment of Orissa super cyclone of 1999, observed that there were 29 cyclone shelters in coastal Orissa, which was one-tenth of the mandated 290 shelters. Similar adaptation deficits were found to exist in other high vulnerable states of Andhra Pradesh, West Bengal, and Gujarat. Based on the past experience on capacity assessment of cyclone shelters the Ministry of Home Affairs observes that at least 60% of the total population in high vulnerable areas could use the cyclone shelters during emergencies. In spite of the recognition of the fact that cyclone shelters reduce damages significantly, the number of multipurpose cyclone shelters in highly vulnerable districts is way too short of that mentioned in prescribed norms.

In this study, we calculate the number of cyclone shelters that could be required in the highly vulnerable districts. High vulnerable areas were identified through historical data based on the frequency and intensity of the cyclones. We also included the present deficit in the number of cyclone shelters (adaptation deficit) in our analysis. Population in high vulnerable areas was projected till 2030, and the number of cyclone shelters with a capacity of 2000 people per shelter was calculated. Total cost for mitigating cyclone risk though multipurpose cyclone shelters was calculated as the product of number of projected cyclone shelters and the per unit cost of a cyclone shelter.[22]

Results

The focus of the overall study is to understand the magnitude of resources required to cope with adverse effects of climate change. The costs given here are only for planned adaptation measures. Annualized costs for each component, considered in the study, are given subsequently. The total adaptation cost in the coastal zones in India, from the present to 2030, is $23.2 billion, which translates to 0.063% of current GDP,[23,24] when the costs are annualized for 30 years. The results of individual components are presented in Table 5.

---

[22] For calculations, refer Annexure V.

[23] Current GDP of the country is $1209.7 billion.

[24] Review of the economy 2008/09. Details available at <http://eac.gov.in/reports/eac_rv0809.pdf>.

Table 5 Results

Adaptation option	Costs in $ billion to 2030	Costs in $ billion/year (annualized over 30 years)
Coastal protection	20.40	0.68
Shelterbelt plantation	2.5	0.083
Cyclone shelters	0.3	0.001
Total	23.2	0.764

Cost of coastal protection varies with SLR. By the end of this century, for about 0.59 m of SLR, which is the worst case scenario projection, the costs are projected to touch $1.5 billion per year as illustrated in Figure 4.

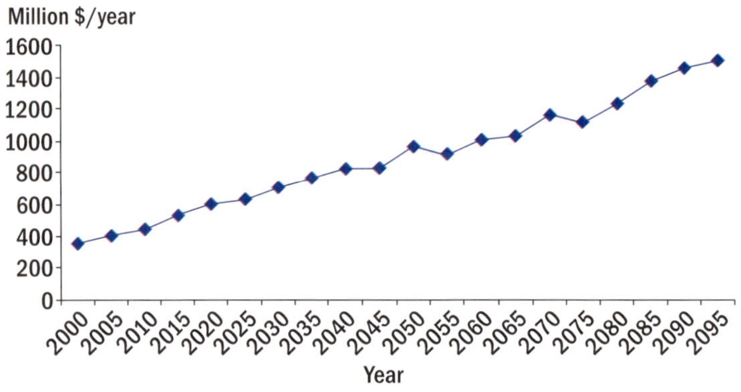

Figure 4 Graph representing the time path of costs for coastal protection

Capital cost for coastal protection would vary with the time frame considered for the analysis; for instance, if a country is planning for 2100, then capital investment would have to be designed keeping the SLR scenario for 2100 in mind. Operation and maintenance costs (variable costs) of coastal protection are very minute compared to the capital costs, and therefore, have not been considered. The exception is variable costs for cyclone shelters, estimated to be 6% of the capital costs and have been accounted in the final estimates. Costs of maintaining and protecting the plantations for five years (after which they can self sustain) have also been included in the estimates.[25]

The costs for shelterbelt plantation and cyclone shelters should be taken as the upper limit. With regard to shelterbelt plantations, there were no other means to establish the spatial distribution and composition of the plantation, which necessitated identifying and adopting wasteland area in the coastal districts for shelterbelt plantation. Cyclone shelters are often used as schools and community halls, and therefore, have a very high degree of co-benefits, which have not been accounted for. Likewise, the benefits of mangrove regeneration are more than just protection from cyclones.

Mangrove regeneration, shelterbelt plantation, and construction of cyclone shelters should be pursued with a sense of urgency, preferably within the next two five-year plans. This is primarily because these strategies are economically efficient no-regret[26] adaptation options. The benefits of these strategies outdo costs by a large margin. They can be implemented within the existing institutions under existing laws.

If we look at some India-specific adaptation cost estimates (Table 6), the estimates have ranged as low as 0.074% of GDP to 0.25% of GDP, depending upon the scenario selected, time frame considered, and assumptions made.

Table 6 Estimates of coastal protection for India

Study	Estimate
IPCC (1990)	0.14% of GDP[a]
Asthana (1994)	0.18% of GDP[b]
ADB (1994)	0.074% of GNP[b]
Deke, Hooss, Kasten, *et al.* (2001)	0.25% of GDP of 1998 levels

[a] costs divided over 100 years; [b] costs divided over 40 years
GDP – gross domestic product; GNP – gross national product

These figures are essentially the costs of protecting the coast for an SLR of 1 m. The table highlights that even for a costly adaptation option like coastal protection (sea dykes), the costs of adaptation are a very small fraction of the GDP.

[25] For detailed break-up of costs from DIVA model, refer Annexure IV.

[26] No-regret adaptation options are measures that could be justified even without anthropogenic climate change.

Knowledge gaps and limitations

Although the study is a significant way forward, lack of quantitative impact assessments and data unavailability limited the scope of the study. Further, the study does not consider many low-cost high impact adaptation options in the analysis. Another significant drawback of this study, as with the earlier studies on estimates of adaptation costs in coastal zones, is its bias towards engineering approaches. There has been growing concern over engineering approaches for coastal protection, and there is an increasing focus on soft approaches. The subsequent section lists some issues that require further analysis.

Need for a better understanding of impacts on the coastal zones

As noted in the earlier section, coastal zones are influenced both by the activities of sea and land. Even in terms of climatic factors impacting coastal systems, we should bear in mind that they can occur singularly or in combinations, traversing various pathways. For example, there could be a scenario of an increase in temperature and an onset of tropical cyclone along with accelerated SLR, each having different intermediate impacts. These changes, coupled with changes in socio-economic dynamics, would complicate the already complex interactions between climate and social systems. Impact modelling studies generally assume a direct cause-and-effect relationship, focused on a limited spatial scale, and do not account for impacts of non-climate variables nor the multiple feedback effects that operate in coupled natural and social systems (Parry and Carter 1998). A better understanding of impacts would enable designing robust adaptation responses and thus better estimates.

Narrow scope of the study

The study does not consider important impacts and ignores several significant cost-effective adaptation measures. National Disaster Management Act, 2005, reports that the death toll from cyclones has reduced in the past decade due largely to improvements in warnings and preparedness, wider public awareness, and a stronger sense of community responsibility. Disaster warning, planning, and management, therefore, should have been part of an important component in the analysis. As also noted earlier, there are rice-growing belts all along the coast. It has been suggested (IPCC 1991) that shift to salt-resistant crop varieties and aquaculture could help reduce the effects of SLR on agriculture.

Ignores second order impacts

Impacts on livelihoods, health, food security, and water availability are not accounted for in the study, as well as the impacts on cultural resources and value.

Coastal megacities and infrastructure

Woodroffe, Nicholls, Saito, *et al.* (2006) and Nicholls (1995) highlight the highly sensitive nature of coastal deltas and megacities with respect to climate change. Coastal flooding would put pressure on already pressed civic amenities and urban infrastructure. Coastal infrastructure, tourist activities, and on-shore oil exploration are at risk from SLR. Under conservative estimates, frequent floods and salt-water intrusion will affect the structural stability of high-rise buildings, which are mushrooming at an increasing rate along the urban areas (Kumar, Jawali, and Tandon 2008). There is a high level of confidence that SLR will increase the cost of protecting infrastructure located in the coastal regions. IIMA (2003) analysis indicates that infrastructure having long life will become more vulnerable to the climatic change 50 years from now. Although the DIVA model assigns extra weight for cities by virtue of considering GDP and population density in assessing optimal coastal protection, it is important to assess costs for measures like realignment and modification of railways, roads, and sewer networks, which are critical for coastal cities to function, and hence, merit consideration in our analysis.

Migration as an adaptive response

Ericson, Vorosmarty, Dingman, *et al.* (2006) in their analysis of global mega deltas, through a generalized modelling approach, demonstrated that most of the population in the deltas is at a very high risk due to coastal erosion and land loss, and 75% of the total affected population is located in the Asian mega deltas. Sundarbans delta, the largest delta in the world, could be the worst affected, forcing millions of people to migrate, with severe adverse trans-boundary implications (Yohe 1990; Alam 2003). It has also been observed that intense cyclonic storms induce migration, as exemplified in the case of Orissa Super Cyclone and Cyclone Aila (Alam 2003). Migration is a form of retreat, one amongst the three broad categories of adaptation measures considered in the study. Migration could be seasonal, temporary or permanent corresponding to partial or complete retreat. The projected increase in frequency and intensity of extreme events along with SLR could force more people to migrate. In order to avoid secondary complications of displacement and migration, it is important that governments plan and manage, which necessitates managed retreat with appropriate institutional arrangements. The study ignores migration, an important component of adaptation in the coastal zone.

Focus on engineering approaches to adaptation

Engineering approaches for coastal protection are likely to be maladaptive in longer time scales. Apart from prohibitive costs that would necessarily be involved, there would be large-scale environmental impacts, including changes in the natural topography of the area and the consequent and as yet incomprehensible impacts on the terrestrial environment, the marine environment, and so on. In general, engineering structures, such as damming, channelization, and diversions of coastal waterways, harden the coast, change circulation patterns, and alter freshwater quality sediment and nutrient delivery (Nicholls, Wong, Burkett, *et al.* 2007). Natural systems are known to be directly or indirectly altered, even by soft engineering solutions, such as beach nourishment and fore-dune construction (Nordstrom 2000; Hamm and Stive 2002). For instance, the state of Kerala has built 354 km of sea walls to prevent sea erosion. Scientists at Thiruvananthapuram, the base Center for Earth Science Studies (CESS),[27] however, observed that sea walls affect the hydrodynamics of the entire coast, causing mass migration of fish. According to them, these structures hinder sediment movement, leading to severe erosion on their leeward side.[28]

---

[27] Details available at <http://www.cessind.org/>.

[28] Details available at <http://www.indiaenvironmentportal.org.in/node/39993>.

Therefore, there is an increased emphasis on soft approaches like beach nourishment, dune management, and shelter belts, and non-structural interventions like capacity building, information sharing, policy regulations, and strategic development (setbacks and zonation), accompanied by suitable institutional arrangements.

Need for bottom-up analysis

A major constraint in analysing soft adaptation approaches is that they are highly site-specific in nature. This mandates a detailed bottom-up analysis to compliment the findings of the top-down modelling studies. This bottom-up information could directly feed into adaptive policy-making.

Conclusion

The study follows an adaptation-centric approach departing from the impact-centric approach of the earlier studies. Although the study follows the conventional approach in assessing the optimal level of coastal protection due to SLR and subsequent costs associated with that, it also makes an attempt to overcome the limitations of that approach by including soft interventions like beach nourishment, shelterbelt plantations, mangrove regeneration, and cyclone risk mitigation in the analysis, which are very important for India, considering its vulnerability to tropical cyclones and the accelerated rate of ecosystem degradation.

Most of the earlier estimates considered an SLR scenario (1 m) and subsequently calculated the costs for protecting the coasts, considering only hard structural interventions. In this respect, this study is a way forward. First, unlike the earlier studies (ADB 1994; Deke, Hooss, Kasten, *et al.* 2001), which considered a static SLR scenario, the starting point of the DIVA model employed in the study is an emission scenario, which is used to compute SLR scenario for a given time frame and the consequent responses. Even under worst case scenario, assessment of the IPCC predicts that the SLR could be about 0.59 m by the end of this century. Second, the model not only considers hard options for coastal protection but also looks at soft adaptation measures like beach nourishment. This is a significant inclusion in the Indian context. For instance, in the case of Goa, the beach is of prime importance from a tourism perspective. With a significant percentage of the state's economy and labour force depending on tourism, construction of sea walls would clearly negate the tourism potential of these beaches.

The figures provided here should be regarded as just indicative and should be aimed at informing policy-makers to make better adaptation decisions. The study ignores quite a few important adaptation options: it ignores co-benefits of adaptation and development. We would reiterate that lack of quantitative impact assessments and data unavailability limited the scope of the study. Nevertheless, it goes further than previous studies in terms of spatial detail and choice of adaptation options considered.

ANNEXURES

Annexure I

Nature and extent of coastal ecosystems in India	
Coastal ecosystem	Extent (km^2)
Tidal/mud flats	23621
Sandy beaches	4210
Mangroves	4445
Coral reefs	2375
Salt marshes	1698
Lagoons	1564
Estuaries	1540
Other vegetation	1391
Aquaculture ponds	769
Salt pans	655
Creeks	192
Rocky coasts	177
Back waters	171
Total	42763

Annexure II

Policies and programmes relevant in the coastal zones.

Policy/ programme	Extreme events	Mangroves/ Corals	Aquaculture	Settlements/ coastal infrastructure	Agriculture	Water quality
Coastal Regulation Zone Notification, 1991	■	■	■	■		
Indian Fisheries Act, 1987		■	■			
Major Port Trust Act, 1963			■			■
Indian Ports Act, 1908						■
Deep Sea Fishing Policy, 1991		■	■			

Continued...

Policies and programmes relevant in the coastal zones *(contd...)*						
Policy/ programme	Extreme events	Mangroves/ Corals	Aquaculture	Settlements/ coastal infrastructure	Agriculture	Water quality
National Environment Tribunal Act, 1995		■	■		■	
The Indian Wildlife (Protection) Act, 1972		■	■			
Environmental Protection Act, 1986		■	■			■
Coast Guard Act, 1978						■
Biological Diversity Act, 2002		■	■			
Environmental Impact Assessment Notification, 1994	■	■	■	■	■	■
Hazardous Waste Management Act, 1989			■			■
National Flood Control Programme- 1953	■	■		■		
Disaster Management Act 2005	■			■		
National Cyclone risk mitigation project, 2004	■	■	■	■	■	
Mangroves for future Project		■				
Integrated Afforestation and Eco-Development Projects Scheme, 2000	■	■				

Note Policies and programmes are mapped with respect to key sectors in the coastal zones.

Annexure III

Physical features of the coastal zones in India considered in DIVA.		
Feature	Unit	Value
Coastal floodplain area	km^2	59 179.47
Coastal floodplain population	thousands	24 445.28
Relative sea level change (since 1995)	M	0.01792626
Total wetland area	km^2	18 597.92
Total wetland monetary value	million $	16.51228
Coastal forest area	km^2	23.42984
Coastal forest monetary value	million $	0.05043604
Mangrove area	km^2	7411.56
Mangrove monetary value	million $	3.000496
Saltmarsh area	km^2	4835.565
Saltmarsh monetary value	million $	0.8792855
Freshwater marsh monetary value	million $	0
Freshwater marsh area	km^2	0
High unvegetated wetland area	km^2	0

Annexure IV

Shelterbelt plantations: calculation	
Wasteland area in the coastal districts as specified by National Wasteland Development Board, Planning Commission of India (2005)	
Andhra Pradesh	Area (in km^2)
Visakhapatnam	2387.28
East Godavari	1449.75
Prakasam	3783.05
Srikakulam	1102.41
Vizianagram	770.27
Nellore	4918.16
Gujarat	
Mahesana	618.87
Bhavnagar	1961.77
Jamnagar	2883.58
Junagadh	3141.14

Continued...

Shelterbelt plantations: calculation *(Contd...)*	
Wasteland area in the coastal districts as specified by National Wasteland Development Board, Planning Commission of India (2005)	
Gujarat	Area (in km^2)
Kheda	526.98
Banaskantha	2742.96
Rajkot	1996.46
Amreli	499.62
Ahmedabad	1099.55
Goa	613.27
Kerala	
Kasaragod	98.14
Malapuram	123.67
Maharashtra	
Ratnagiri	3453.94
Thane	2419.52
Orissa	
Cuttack	768.53
Puri	1751.85
Balasore	218.42
Ganjam	4310.84
Tamil Nadu	
Tiruchirapalli	1389.5
Pudukkottai	1147.29
South Arcot	1433.95
Kanyakumari	216.52
Ramanathapuram	667.2
Total	48 494.49 In hectares = 4 849 500 ha
Cost of raising 1 hectare (ha) of shelterbelts given by the Ministry of Environment and Forests = Rs 24 596 Total costs for coastal shelterbelts is found to be Rs 119 278 302 000/11.92 billion (and annualized cost is Rs 37 billion per year).	

Annexure V

Cyclone shelters: calculation		
Parameter	Baseline	Climate change
Total population	4 125 412	6 426 062
Population needing cyclone shelters (60%)	2 475 247	3 855 637.2
Number of cyclone shelters	1238	1928
Unit Cost	7 064 350	7 064 350
Capital cost	8 745 665 300	13 620 066 800
Operation and maintenance cost (6% of capital investment)	423 858	8 901 018
Total costs	8.8 billion	13.63 billion

Annexure VI

Outputs of DIVA		
Item	Unit	Figure
Total adaptation costs	million $/year	679.7488
Beach nourishment costs	million $/year	32.11699
Total residual damage costs	million $/year	311.3651
Land loss costs	million $/year	0.260562
People actually flooded	thousand/year	466.7247
Loss of flats	m^2/year	1 043 963
Tidal basin demand for nourishment	m^3/year	58 516 920
Net loss of wetland area	km^2	1593.183
People actually flooded	thousand/year	466.7247
Coastal floodplain area	km^2	61390.88
Coastal floodplain population	Thousand	39 697.93
River dike costs	million $/year	1.991398
River flood costs	million $/year	0.030446
Relative sea level change (since 1995)	M	0.173573
Salinity intrusion costs	million $/year	1.356598
Sand loss due to tidal basins	m^3/year	1 456 760
Sand loss total	m^3/year	19 030 660
Sea dike costs	million $/year	595.3792
Sea flood costs	million $/year	309.7175

Continued...

Outputs of DIVA *(Contd...)*		
Items	Unit	Figure
Total wetland area	km^2	17 171.66
Coastal forest area	km^2	22.35052
Coastal forest monetary value	million $	0.138062
Total wetland monetary value	million $	43.67961
Wetland nourishment costs	million $/year	0.261321
Low unvegetated wetlands area	km^2	5665.055
Low unvegetated wetland monetary value	million $	33.18657
Mangrove area	km^2	6972.085
Mangrove monetary value	million $	8.013652
Saltmarsh area	km^2	4512.169
Saltmarsh monetary value	million $	2.341317

References

ADB (Asian Development Bank). 1994. **Climate Change in Asia**. [India Country Report, Manila, Philippines]

Adger W N, Agrawala S, Mirza M M Q, Conde C, O'Brien K, Pulhin J, Pulwarty R, Smit B, Takahashi K. 2007. **Assessment of adaptation practices, options, constraints and capacity**. In *Climate Change 2007: Impacts, Adaptation, and Vulnerability. Contribution of Working Group II to the Fourth Assessment Report of the Intergovernmental Panel on Climate Change*, pp. 717–743, edited by M L O F Parry, J P Canziani, P J Palutikof, van der Linden, and C E Hanson. Cambridge, UK: Cambridge University Press

ADPC (Asian Disaster Preparedness Centre). 2005. **Country paper for India**. [Paper presented at the *International Workshop on Strengthening the Resilience of Local Communities in Coastal Areas*, Copenhagen, 16–18 November 2005]

Alam S 2003. **Environmentally induced migration from Bangladesh to India.** *Strategic Analysis* **27**(3): 21–25

Asthana V. 1994. ***Impacts of Greenhouse-induced Sea Level Rise on the Islands and Coasts of India***. New Delhi: Jawaharlal Nehru University

Badola R and Hussain S H. 2005. **Valuing ecosystem functions: an empirical study on the storm protection function of Bhitarkanika mangrove ecosystem, India**. *Environmental Conservation* **32**: 1–8

Balaji S. 2005. **Bioshield for coastal protection: mangroves, shelterbelts and coral reefs**. In *National Workshop on Tsunami Effects and Mitigation Measures*, pp. 153–159, edited by V Sundar. Chennai: Department of Ocean Engineering

Bijlsma L, Ehler C N, Klein R J T, Julshrestha S M, McLean R F, Mimura N. 1996. **Coastal zones and small islands**. In *Climate Change 1995: Impacts, Adaptations, and Mitigation of Climate Change: scientific-technical analysis. Contribution of Working Group II to the Second Assessment Report of the Intergovernmental Panel on Climate Change*, edited by R T Watson, M C Zinyowera, and R H Moss. Cambridge: Cambridge University Press. 879 pp.

Blasco F and Aizpuru M .2002. **Mangroves along the coastal stretch of Bay of Bengal: present status**. *Indian Journal of Marine Sciences* **31**(1): 9–20

Bruun P. 1962. **Sea-level rise as a cause of shore erosion**. *Journal Waterways and Harbours Division (ASCE)* **88**: 117–130

Das S. 2007. **Mangroves provide protection during cyclone: evidences from the super cyclone of October 1999 in India**. Details available at <http://www.feem-web.it/ess/files/das.pdf>, last accessed on 23 January 2010

Das S and Vincent J R. 2009. **Mangroves protected villages and reduced death toll during Indian super cyclone**. *Proceedings of the National Academy Science* **106:** 7357–7360

Deke O, Hooss K G, Kasten C, Klepper G, Springer K. 2001. **Economic impact of climate change: simulations with a regionalized climate-economy model**. [Kiel Working Paper No. 1065, Kiel Institute of World Economics]

Erratapuzha J J and George V. 1980. ***Geology and Geomorphology of Kerala***. 83 pp. [Geological Survey of India Special Publication]

EEA (European Environment Agency). 2007. **Climate change: the cost of inaction and the cost of adaptation**. [EEA Technical Report No. 13/2007]. Details available at <http://www.eea.europa.eu/publications/technical_report_2007_13/at_download>, last accessed on 23 September 2009

ENVIS (Environmental Information System) Centre. 2007. **Global warming...Melting ice: a hot topic?** *ENVIS Newsletter on State of Environment* 4(2) [ENVIS Centre, Department of Environment, Government of Tamil Nadu]. Details available at <www.tnenvis.nic.in/images/junenl07.pdf>

Ericson J P, Vorosmarty C J, Dingman S L, Ward L G, Meybeck M. 2006. **Effective sea-level rise and deltas: causes of change and human dimension implications**. *Global Planet Change* **50**: 63–82

Fankhauser S. 1994. **Protection versus retreat: the economic costs of sea level rise**. *Environment and Planning* **A27**: 299–319

Field C D. 1995. **Impacts of expected climate change on mangroves**. *Hydrobiologia* **295**(1–3): 75–81

GoI (Government of India). 1981. **Report on development of coastal areas affected by salinity planning commission**. New Delhi: National Committee on the Development of Backward Areas

GoI (Government of India). 1997. **Government of India reporting to UN commission on sustainable development**. Details available at <http://www.un.org/esa/earthsummit/india-cp.htm#chap17>, last accessed on 3 November 2009

GoI (Government of India). 2004. ***India's Initial National Communication to the United Nations Framework Convention on Climate Change.*** New Delhi: Ministry of Environment and Forests, GoI

Hamm L and Stive M (eds). 2002. **Shore Nourishment in Europe, Co. Eng.,** A Special Issue **47**: 2. 264 pp.

Henderson-Sellers A, Zhang H, Berz G, Emanuel K, Gray W, Landsea C, Holland G, Lighthill J, Shieh S L, Webster P, McGuffie K. 1998. **Tropical cyclones and global climate change: a post-IPCC assessment**. *Bulletin of the American Meteorological Society* **79**: 19–38

Hoozemans F M J, Marchand M, and Pennekamp H. 1993. **Sea level rise: a global vulnerability assessment-vulnerability assessment for population, coastal wetlands, and rice production on a global scale.** In *Delft Hydraulics,* 2nd edn The Netherlands: Rijkswaterstaat. 184 pp.

IIMA (Indian Institute of Management Ahmedabad). 2003. **Climate change impacts on infrastructure.** In *Development and Climate: an assessment for India*, pp. 65–76, edited by P R Shukla, R Nair, M Kapse, A Garg, S Balasubramaniam, D Menon, and K K Sharma. Ahmedabad: IIMA

IPCC (Intergovernmental Panel on Climate Change). 1990. ***Strategies for Adaptation to Sea Level Rise. IPCC Response Strategies Working Group***. The Hague, the Netherlands: Ministry of Transport and Public Works and Ministry of Housing. 122 pp.

IPCC (Intergovernmental Panel on Climate Change). 1991. ***Climate change: the IPCC response strategies***. Washington, DC: Island Press

IPCC (Intergovernmental Panel on Climate Change). 1992. **Global Climate Change and the Rising Challenge of Sea**. Report of the Coastal Zone Management Subgroup of the IPCC Response Strategies Working Group. Washington, DC: Island Press

IPCC (Intergovernmental Panel on Climate Change). 2007. **Climate Change 2007: Impacts, Adaptation and Vulnerability. Contribution of Working Group II to the Fourth Assessment Report of the Intergovernmental Panel on Climate Change**, edited by M L Parry, O F Canziani, J P Palutikof, P J van der Linden, C E Hanson. Cambridge, UK: Cambridge University Press. 976 pp.

Jagtap T G, Kulkarni V A, and Verlekar X N. 2008. **Vulnerability and adaptation of ecologically sensitive mangrove habitats to the changing climate.** In *Proceedings of Conference on Marine Problems and Specific Solutions*, COMPASS 2008, Maldives 15–18 June 2008. Chennai: National Institute of Oceanography

Kumar R, Jawali P, and Tandon S. 2008. **Economic impact of climate change on Mumbai, India**. *Regional Health Forum* **12**(1): 38-42

Lal. 2001. **Tropical cyclones in a warmer world.** *Current Science* **80**(*9*): *1103–1104*

Leatherman and Nicholls. 1995. **Accelerated sea-level rise and developed countries: an overview.** *Journal of Coastal Research,* Special Issue **14**: 1–14

Mani J S. 2000. **Coastal processes in Tamil Nadu and Kerala: environmental and socio-economic aspects**. In *Environmental Problems of Coastal Areas in India*, pp. 201–216, edited by V K Sharma. New Delhi: Bookwell Press.

Massel S R. 1999. **Tides and waves in mangrove forests**. In *Fluid Mechanics for Marine Ecologists*, pp. 418–425. Germany: Springer-Verlag

Mazda Y, Magi M, Kogo M, Hong P N. 1997. **Mangroves as a coastal protection from waves in the Tong Kong Delta, Vietnam**. *Mangroves and Salt Marshes* **1**: 127–135

McFadden L, Vafeidis A, and Nicholls R J. 2003. **A coastal database for global impact and vulnerability analysis**. In *Proceedings of Coastal Sediments 03*. New York: American Society of Civil Engineers.

McFadden L, Nichollos R J, Vafiedis A T, and Tol R S J. 2007. **A methodology for modelling coastal space for global assessment**. *Journal of Coastal Research,* **23**(4): 911-920.

McLean R, Tsyban A, Burkett V, Codignotto J O, Forbes D L, Mimura N, Beamish R J, Ittekkot V. 2001. **Coastal zone and marine ecosystems**. In *Climate Change 2001: Impacts, Adaptation and Vulnerability*, pp. 343-380. Contribution of Working Group II to the Third Assessment Report of the Intergovernmental Panel on Climate Change, edited by J J McCarthy, O F Canziani, N A Leary, D J Dokken, and K S White. Cambridge: Cambridge University Press

Mendelsohn R. 2000. **Efficient adaptation to climate change**. *Climatic Change* **45**: 583–600

Mendelsohn R, Morrison W, Schlesinger M E, Andronova N G. 2000. **Country-specific market impacts of climate change**. *Climatic Change* **45**: 553–569

Mitra A P. 2004. **Climate in South Asia and water resources: introductory note**. In *Climate Change and Water Resources in South Asia*, edited by M Amir. [Proceedings of Year End Workshop]

MoEF (Ministry of Environment and Forests). 2002. ***Annual Report 2001/02***. Details available at <http://envfor.nic.in/report/0102/chap06.html>, last accessed on 21 November 2009.

MoEF (Ministry of Environment and Forests). 2007. ***Annual Report 2006/07***. New Delhi: MoEF, Government of India

NDMA (National Disaster Management Authority). 2008. **National guidelines for the management of cyclones**. Details available at <http://ndma.gov.in/ndma/guidelines/Cyclones.pdf>, last accessed on 3 April 2009

Ng W S and Mendelsohn R. 2005. **The impact of sea level rise on Singapore**. *Environment Development Economics* **10**: 210–215

Nicholls R J. 1995. **Coastal megacities and climate change**. *GeoJournal* **37**(3): 369–379

Nicholls R J and Hoozemans F M J, 1996. **The Mediterranean: vulnerability to coastal implications of climate change**. *Ocean and Coastal Management* **31**: 105–132

Nicholls R J and Mimura N. 1998. **Regional issues raised by sea level rise and their policy implications**. *Climate Research* **11**: 5–18

Nicholls R J. 2000. **An analysis of the flood implications of the IPCC second assessment global sea-level rise scenarios**. In *Floods*, pp. 148–162, edited by D J Parker. London: Routledge

Nicholls R J and Small C. 2002. **Improved estimates of coastal population and exposure to hazards released**. *EOS Transactions*, **83**(2), 301 and 305

Nicholls R J and Lowe J A. 2004. **Benefits of mitigation of climate change for coastal areas**. *Global Environment Change* **14**: 229–244

Nicholls R J and Tol R S J. 2006. **Impacts and responses to sea-level rise: a global analysis of the SRES scenarios over the 21st century**. *Philosophical Transactions of the Royal Society* **A364**: 1073–1095

Nicholls R J, Wong P P, Burkett V R, Codignotto J O, Hay J E, McLean R F, Ragoonaden S, Woodroffe C D. 2007. **Coastal systems and low-lying areas**. In *Climate Change 2007: impacts, adaptation and vulnerability*, pp. 315–356, edited by M L Parry, O F Canziani, J P Palutikof, P J van der Linden, and C E Hanson. Cambridge, UK: Cambridge University Press. [Contribution of Working Group II to the *Fourth Assessment Report* of the Intergovernmental Panel on Climate Change]

Norahna L, Choudri B S, and Nairy K S. 2003. **Relative vulnerability of the coastal districts to a potential sea level rise along the coastline of India.** In *Environmental Threats, Vulnerability, and Adaptation: case studies*, pp. 121–143. New Delhi: TERI

Nordstrom K F. 2000. ***Beaches and Dunes of Developed Coasts***. Cambridge: Cambridge University Press. 338 pp.

Patwardhan A and Narayanan K. 2003. **Assessment of vulnerability of Indian Coastal Zones to Climate Change**. [Paper presented at the *Conference on Water Resources, Coastal Zones, and Human Health*, July 2003, Ministry of Environment and Forests, Government of India]

Patwardhan A, Narayanan K, Parthasarathy D, Sharma U. 2003. **Assesment of vulnerability and adaptive capacity in coastal zones**. In *Climate Change and India: vulnerability assessment and adaptation*, edited by P R Shukla, S K Sharma, N H Ravindranath, A Garg, S Bhattacharya. Hyderabad: Universities Press

Parry M and Carter T. 1998. ***Climate Impact and Adaptation Assessment. A Guide to the IPCC Approach***. London, UK: Earthscan Publications Ltd. 166 pp.

Pielke. 2007. **Future economic damage from tropical cyclones: sensitivities to societal and climate changes.** In *Philosophical Transactions of The Royal Society*. Details available at <http://sciencepolicy.

colorado.edu/admin/publication_files/resource-2517-2007.14.pdf>, last accessed on 21 July 2009

Planning Commission. 2005. ***National Wasteland Development Report.*** New Delhi: Planning Commission, Government of India

Ramesh R and Ramachandran P. 2007. **ENVIS–MoEF on Coastal Zone Management and Coastal Shelter belt**, vol 6. 25 pp.

Selvam T, Ravishankar V M, Karunagaran R, Ramasubramanian P, Eganathan P, Parida A K. 2005. ***Toolkit for Establishing Coastal Bio-shield***. Chennai: M S Swaminathan Research Foundation. 117 pp.

Shetye S R, Gouveia A D, Pathak M C, 1990. **Vulnerability of the Indian coastal region to the damage from sea level rise**. *Current Science* **59**(3): 152–156

Shrestha M L (ed.). 1998. ***The Impact of Tropical Cyclones on the Coastal Regions of SAARC Countries and their Influence in the Region***, p. 14. Agargaon, Bangladesh: SAARC Meteorological Research Centre. 329 pp.

Smith J B, Tol R S J, Ragland S, Fankhauser S, 1998. **Proactive Adaptations to Climate Change: three case studies on infrastructure investments**. [IVM Discussion Paper D98/03] Amsterdam, the Netherlands: Institute for Environmental Studies, Vrije Universiteit, Amsterdam: The Netherlands, pp.14

Stive M. 2004. **How important is global warming for coastal erosion?** *Climatic Change* **64**: 27–39

TERI (The Energy and Resources Institute). 1996. ***The Economic Impact of a One Metre Sea Level Rise on the Indian Coastline: method and case studies***. New Delhi: TERI. [Report No. 93/GW/52]

Titus J, Park R, Leatherman S, Weggle J, Greene M, Brown S, Gaunt C, Trehan M, Yohe G. 1992. **Greenhouse effect and sea level rise: the cost of holding back the sea**. *Coastal Management* **19**: 219–233

Tol R S J. 1995. **The damage costs of climate change: towards more comprehensive calculations**. *Environmental and Resource Economics* **5**: 353–374

Tol R S J. 2002. **Estimates of the damage costs of climate change. Part 1: benchmark estimates**. *Environment and Resource Economics* **21**: 47–73

Tol R S J, Bohn M, Downing T E, Guillerminet M L, Hizsnyik E, Kasperson R, Lonsdale K, Mays C. 2006. **Adaptation to five metres of sea-level rise**. *Journal of Risk Research* **9**: 467–482

Toman M. 2006. **Values in the economics of climate change**. *Environmental Values* **15:** 365–379

Unnikrishnan A S, Rupa Kumar K, Fernandes S E, Michael G S, Patwardhan S K. 2006. **Sea level changes along the Indian coast: observations and projections**. *Current Science* **90**: 362–368

US Commission on Marine Science, Engineering, and Resources. 1969. ***Our Nation and the Sea: a plan for national action: report of the commission on marine science, engineering and resources***. Washington, DC: US Commission on Marine Science, Engineering, and Resources

Vafeidis A T, Nicholls R J, McFadden L, Tol R S J, Hinkel J, Spencer T, Grashoff P S, Boot G, Klein R J T. 2008. **A new global coastal database for impact and vulnerability analysis to sea-level rise.** *Journal of Coastal Research* **24**(4): 917–924

Woodroffe C D, Nicholls R J, Saito Y, Chen Z, Goodbred S L. 2006. **Landscape variability and the response of Asian megadeltas to environmental change**. In *Global Change and Integrated Coastal Management: the Asia-Pacific region*, pp. 277–314, edited by N Harvey. New York: Springer

Yohe G W and Schlesinger M E. 1998. **Sea-level change: the expected economic cost of protection or abandonment in the United States**. *Climatic Change* **38**: 447–472

Yohe G. 1990. **The cost of not holding back the sea**. *Coast Management* **18**: 403–431

Yohe G W, Neumann J E, and Ameden H. 1995. **Assessing the economic cost of greenhouse-induced sea level rise: methods and applications in support of a national survey.** *Journal of Environmental Economics and Management* **29**: S-78–S-97

Zhang K, Douglas B, and Leatherman S. 2004. **Global warming and coastal erosion**. *Climatic Change* **64**: 41–58

six

Water resources

Sambita Ghosh,[a] Suruchi Bhadwal,[b] and Julia Martin Ortega[c]

Background

India receives about 4000 billion cubic metres (BCM) of rain each year, of which about 80% of the rainfall is received mainly during the south-west monsoon season, from June to September. Also the rainfall is highly variable and non-uniformly spread across various regions, ranging from 150 mm a year in Rajasthan to 11 690 mm a year in Meghalaya, with snowfall restricted to the Himalayan region.

Out of the total precipitation of about 4000 BCM a year, including snowfall, surface water and groundwater availability is estimated to be 1869 BCM. However, all this water cannot be put to beneficial use owing to various constraints of topography and the form in which water is stored. Out of 1869 BCM of water, only about 1123 BCM can be used, of which 690 BCM is available from surface water sources and 433 BCM from groundwater sources (NATCOM 2004).

India has a river system comprising 20 river basins (Annexure I), of which 12 can be categorized as major river basins, with a catchment area >20 000 km^2. The basins are classified as the Himalayan rivers, peninsular rivers, coastal rivers, and rivers of inland drainage basin. The Himalayan rivers are snow-fed and are perennial. Rivers like the Ganga, Brahmaputra, and Indus originate from the Himalayas

---

[a] Research Associate, Centre for Global Environment Research, Earth Science and Climate Change Division, The Energy and Resources Institute (TERI), New Delhi, India.

[b] Fellow, Centre for Global Environment Research, Earth Science and Climate Change Division, The Energy and Resources Institute (TERI), New Delhi, India.

[c] Researcher, Basque Centre for Climate Change(BC3), Bilbao, Spain.

and carry water throughout the year. The snow and ice melt of the Himalayas and the base flow contribute to the flows during the lean season. Average water yield per unit area of the Himalayan rivers is almost twice that of the south peninsular river system, indicating the importance of snow- and glacier-fed rivers. The peninsular rivers are rain-fed and, therefore, vary in volume during the year. Vast areas of the country are affected by floods and droughts. While a third of the country is drought prone, floods affect an average area of 9 million hectares (Mha) per year, and the area susceptible to floods is about 40 Mha. The area affected by floods has also been found to have increased over the years, from 19 Mha in 1953 to 40 Mha in 2003 (World Bank 2009), affecting nearly 33 million people. At the same time, as many as 99 districts, spread over 14 states, were identified by the Central Water Commission (CWC 2005) as drought prone. Most of the drought-prone areas are concentrated in the states of Rajasthan, Karnataka, Andhra Pradesh, Maharashtra, and Gujarat.

Apart from the various rivers of the country, groundwater is also an important source for drinking, irrigation, industrial uses, and so on. Rainfall is the major source of groundwater recharge, which is also supplemented by other sources such as recharge from canals, irrigated fields, and surface waterbodies. The annual potential of natural groundwater recharge from rainfall in India is about 343 BCM, which is 8.56% of the total annual rainfall of the country. The annual potential of groundwater recharge augmentation from canal irrigation system is about 90 BCM. The total annual replenishable groundwater resources of the country have been assessed to be 433 BCM, and the net annual ground water availability is estimated to be 399 BCM. A quantum of 34 BCM is set aside for natural discharge during the non-monsoon period for maintaining flows in springs, rivers, and streams. Due to heavy extraction of groundwater and limited recharge, groundwater depletion has markedly increased in most dryland regions of the country.

Climate change impacts on water resources and costs of adaptation: a literature survey

Impacts on freshwater resources

The impacts of climate change on freshwater systems include a combination of factors comprising increase in temperature and sea level, and precipitation variability (Kundzewicz, Mata, Arnell, *et al.* 2007), which are predicted by the Integovernmental Panel on Climate Change (IPCC) with a high to very high level of confidence (IPCC 1998). Climate change also challenges existing water resource practices by adding additional uncertainty. Changing precipitation patterns would affect the replenishment of rivers by reducing their flows during the critical summer months when the demand is the highest. Water quality would also emerge as an important issue in this context, which has the potential to endanger human health. Reduced flows in rivers would affect the dilution capacity of pollutants, thus affecting the water quality and further reducing usable sources. Increased glacial melt due to warming is also predicted to affect river flows. As stated in the IPCC reports, glacial melt is expected to increase under changed climate conditions, which is likely to lead to increased summer flows initially and a reduction in the flows as the glaciers reduce in mass. Sea level rise in coastal areas would inundate land areas and displace population living in these areas (see "Coastal zones" chapter for more details). It would also result in an increased risk to life and property. Low-lying coastal cities would be affected and would be at the forefront of impacts. Salt-water intrusion in underground aquifers will be an additional cause of concern in these regions.

Some of the existing literature on climate change impacts in Asia indicates that the rainfall on the continent is most likely to increase, although regional variations would exist. Also, the impacts on glaciers are likely to result in increased snow melt, further contributing to an overall increase in the flows of river systems, which in the short-term might result in an increased risk of flooding, erosion, landslides, and would in the long-term contribute to reduced flows. There is a likelihood that nearly 1 billion people would be affected and would face water shortages (Christensen, Hewitson, Busuioc, *et al.* 2007; Cruz, Harasawa, Lal, *et al.* 2007). The Ganga river is expected to lose two-thirds of its July to September flow, affecting one-third of India's irrigated land and causing water shortages for more than 500 million people in South Asia. The river systems of the Brahamaputra and the Indus are also likely to be affected. Impacts on deltaic regions and degradation of wetlands and ecosystems are likely to be quite prominent. Sea level rise is likely to affect freshwater systems in coastal areas due to salt-water intrusion and contamination of groundwater aquifers.

Adaptation strategies and the costs of adaptation

In order to reduce the negative impacts of climate change, implementing adaptation measures is critical. In the case of freshwater, precipitation variability is likely to put pressure on the need for policy and planning, including infrastructure investments that help address the risks. Most of the adaptation measures reported in the literature for freshwater so far include planned adaptation measures, which require a deliberate policy decision and investment.

The adaptation measures can be both preventive and reactive. Preventive actions aim at avoiding damage due to climate change (example, avoiding restrictions in water supply, preventing the rivers from overflowing, and implementing advisory measures to prevent the damage in case the river does overflow). Reactive actions, on the other hand, help deal with the damage once it has occurred (example, salvation and evacuation of people in case of flooding). Some such measures include re-creation of employment options after drought/floods, capacity building of local authorities, and assistance to small subsistence farmers (EEA 2007). Table 1 provides a summary of adaptation measures based on the type of impacts and response needed.

Table 1 Classification of adaptation measures in the freshwater sector

Impact	Adaptation measure
Decrease in water availability and risk of drought	**Preventive**
	• Promotion of changes in water demand and efficiency of water use. • Establishing pricing and market mechanisms. • Promoting shift water usage within and between sectors (example, agriculture to urban) and sharing/rationing. • Implementing programmes for increasing efficiency (irrigation, industry, and households).

Continued...

Total cost of drainage development = Unit cost of drainage development × Total area covered by basins that are likely to experience a high surplus in water availability ...(10)

Methodology for channel development

Channelling of river streams helps in abating the possibility of floods, following the principle that a sufficiently large and deep channel would lead to the occurrence of floods. The unit cost of channel development was calculated by dividing the total amount of money spent for channel development with the total length of the river.

The total length along which channel development activities have to be undertaken was calculated by adding the lengths of the flood-prone rivers, namely, lower Ganga belt, Mahanadi, and Brahmani and Batarani. To arrive at the total cost of channel development, the unit cost of channel development was multiplied with the total length that requires channel development activities.

Total cost of channel development = Unit cost of channel development × Total length requiring channel development ...(11)

The final figure obtained would be an upper bound as development of channels is already being undertaken in some areas, and all stretches may not require channelization.

Results

The costs indicated in Table 5 are for the time period up to 2050, and these could be annualized assuming that one could distribute the costs from 2010 to 2050. Costs disaggregated at the level of river basins have been provided in Annexure II. The annual costs would be about $0.6 billion.

Table 5 Summary of costs to address water deficit and inland flooding

Adaptation options	Cost (billion $)
Water availability deficit	
Supply side measures	
Rainwater harvesting	0.54

Continued...

Table 5 *Contd...*	
Adaptation options	Cost (billion $)
Desalination of seawater	20.400
Waste water treatment and reuse (irrigation)	0.110
Demand side measures	
Water metering	0.662
Use of water-saving devices (in Rajasthan and 10% Madhya Pradesh)	0.001
Inland flooding	
Structural measures	
Channel improvement	1.420
Drainage development	1.140
Total	24.273

Conclusion

Although some river basins are likely to experience an increase in the run-off, others will face huge deficits. These estimations have been based on impact studies carried out for 12 major basins by Gosain, Rao, and Basuray (2006), as part of India's first national communications. The total deficit was estimated to be about 38.05 BCM, covering the basins in the states of Tamil Nadu, Karnataka, Kerala, Maharashtra, Andhra Pradesh, Gujarat, Rajasthan, and parts of Madhya Pradesh. To be able to address these deficits, both demand and supply side measures were explored. The costing to meet this deficit has been estimated, giving preference for supply side measures vis-à-vis demand side ones, but at the same time not ignoring demand side measures. Within the supply side measures, prioritization of options has been done based on the least cost options (as per the unit cost) and based on the extent to which these options can be applied. While moving from one option to the other, from the least cost option to rather more expensive options, the remaining deficit had been considered, basically indicating that a preference has been given to the least cost options in the estimations. The remaining deficit that could not be covered under the supply side measures was then addressed by considering demand side measures (Figure 1).

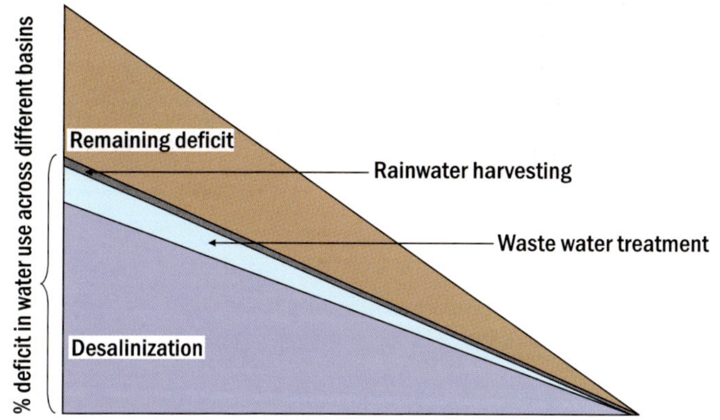

Figure 1 Different measures to address water deficit
Note Supply side measures – 58%; demand side measures and other actions – 42%

Of the various options, RWH can help address only 1.2% of the total deficit, even after all urban rooftop areas would have been considered in the deficit basins. Waste water treatment could help address another 6.1% of the deficit, since the waste water is generated from each of the deficit states. Desalinization as an option then has to be explored given the circumstances, which can help address about 50% of the deficit, given that it has been considered viable for adoption in the coastal states.

The remaining 42% of the deficit can be addressed either through demand side measures, including the use of water-saving devices and metered connections, or through other supply side measures, which have not been quantified in this chapter but addressed in the chapter on agriculture, and which include artificial recharge; reduce, reuse, recycle activities; and so on or by taking into account the demand side measures. Although demand side measures have an important role to play, in our view, they cannot fill the whole of the remaining gap. We conclude, therefore, that despite the demand side initiatives, there would remain a deficit of 4.6 BCM, which would be needed to be addressed. This analysis in no way is holistic, given that the shares under each option can vary and the underlying assumptions considered, and therefore, one should be careful while interpreting the results.

Our cost estimate of \$0.6 billion per annum is quite modest compared with, for example, the World Bank study for South Asia as a whole, which indicates annual costs ranging between \$2.9 billion and \$7.6 billion for water supply and flood management. Considering India as constituting about 73% of South Asia in terms of population, these would imply costs between \$2.0 billion and 5.4 billion per annum from 2010 to 2050. We would attribute this to two factors. First, we have included a number of low-cost demand options to fill the deficit. Second, we have not included measures to fill the whole deficit: as noted above, about 4.6 BCM of deficit remains to be accounted for. In addition, we have not accounted for all measures needed to address inland flooding. Large-scale damages for which strategies have not been explored include those concerning land use change/floodplains and relocation. Disaster preparedness measures, including early warning systems/forecasting systems, communication campaigns, evacuation plans and first aid, immediate shelter provision, and temporary population relocation, have also not been considered. Interlinking of rivers as an option the associated costs have also not been included. However, this would have been logical, given that certain areas are projected to observe a surplus in run-off compared with a large proportion of area in the country experiencing huge deficits. Through published sources, the cost for such a programme has been indicated to be about \$125 billion and could be explored to meet deficits in certain river systems (IWMI 2009).

In addition to the above, we need to recognize that preventive measures will surely not be sufficient to avoid the damage totally. Thus, despite taking preventive measures, there would still be people affected by flooding. The effects can include wounds, outbreaks of infectious diseases such as cholera, typhoid, and diarrhoeal diseases that can spread in contaminated floodwater (McMichael, Campbell-Lendrum, Kovats, *et al.* 2004), and outbreak of vector-borne diseases such as leptospirosis (Kríz, Benes, Cástková, *et al.* 1998), as well as psychological morbidity, and non-human damage (economic losses, loss of houses, damage to infrastructure, and so on). The costs of the reactive measures towards combating these effects should have been included more in the form of impacts. However, the chapter as of now has not included these costs, including the treatment cost for sick people due to water-related disease outbreaks in the context of the health sector chapter. With regard to non-human damages, the costs for protection have not been computed. To conclude, there is immense potential to continue the ongoing work and develop more detailed estimations on the costs for adaptation in the water sector in India.

ANNEXURE

Annexure I
Map depicting the river basins in the country

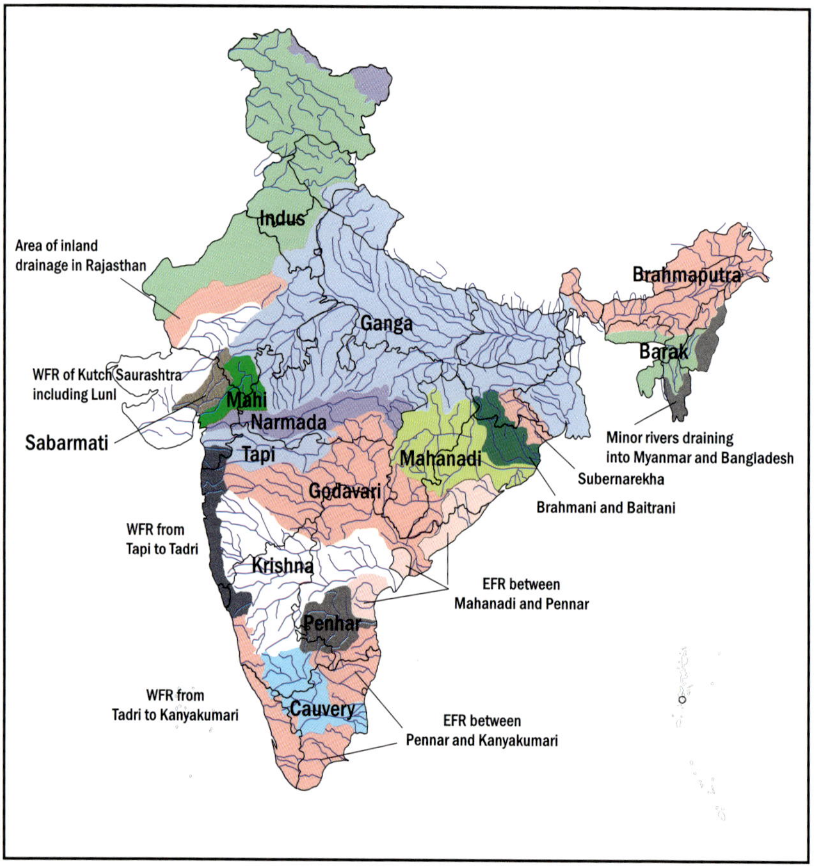

WFR – West flowing river; EFR – east flowing river

Source National Action Plan on Climate Change (NAPCC). 2008. Comprehensive Mission Document (Vol. II).

Annexure II
Results, by basin

Table I. Costs of adaptation measures for water deficit							
River basins facing deficit	Water deficit covered (BCM)			Costs (billion $)			Total costs (billion $)
	RWH	WWT	Desali-nation	RWH	WWT	Desali-nation	
Cauvery	0.143	0.268	0.135	0.08	0.01	0.14	0.23
Krishna	0.150	0.598	9.81	0.11	0.03	10.46	10.59
Mahi	0.009	0.369	0.69	0.09	0.02	1.73	0.84
Pennar	0.140	0.335	2.76	0.09	0.02	2.94	3.04
Sabarmati	0.008	0.369	4.195	0.09	0.02	4.47	4.59
Luni	0.010	0.375	0.81	0.09	0.02	0.86	0.97
Total	0.468	2.314	18.39	0.54	0.11	19.62	20.28

RWH – rainwater harvesting, WWT – waste water treatment

Table II. Channel development

River basins expected to flood in the future	Cost (billion $)
Ganga	0.50
Brahmani and Baitarani	0.54
Mahanadi	0.39
Total	1.42

Table III. Drainage development

River basins expected to flood in the future	Cost (billion $)
Ganga	0.72
Brahmani and Baitarani	0.04
Mahanadi	0.12
Godavari	0.26
Total	1.14

References

Agrawala S and Frankhauser S (eds). 2008. ***Economic Aspects of Adaptation to Climate Change.*** Paris: Organisation for Economic Cooperation and Development (OECD). 133 pp.

Arnell N. 2009. **Costs of adaptation in the water sector**. In *Assessing the Costs of Adaptation to Climate Change: a review of the UNFCCC and other recent estimates*, edited by M Parry, N Arnell, P Berry, D Dodman, S Frankhauser, C Hope, S Kovats, R Nicholls, D Satterhwaite, R Tiffin, T Wheeler *et al.* London: International Institute for Environment and Development and Grantham Institute for Climate Change

Callaway J M, Louw D B, Knomo J C, Hellmuth M E, Sparks D A. 2007. **The Berg River Dynamic Spatial Equilibrium Model: a new tool for assessing the benefits and costs of alternative coping with water demand growth, climate variability and climate change in Wester Cape.** [AIACC Working Paper no. 31]. Details available at <www.aiaccproject.org>

Central Ground Water Board. 2002. **Master plan for artificial recharge to groundwater in India. Drinking water.** Details available at <http://www.nih.ernet.in/nih_rbis/india_information/drinking.htm>, last accessed on 7 January 2010

Christensen J H, Hewitson B, Busuioc A, Chen A, Gao X, Held I, Jones R, Kolli R K, Kwon W T, Laprise R, Magaña Rueda V, Mearns L, Menéndez C G, Räisänen J, Rinke A, Sarr A, Whetton P. 2007. **Regional climate projections**. In *Climate Change 2007: the Physical Science Basis. Contribution of Working Group I to the Fourth Assessment Report of the Intergovernmental Panel on Climate Change*. Cambridge, UK and New York, USA: Cambridge University Press

CPCB (Central Pollution Control Board). 2001. ***Status of Water Supply, Wastewater Generation and Treatment in Class I Cities and Class II towns of India***. CPCB, Ministry of Environment and Forests, Government of India

Cruz R V, Harasawa H, Lal M, Wu S, Anokhin Y, Punsalmaa B, Honda Y, Jafari M, Li C, Huu Ninh N. 2007. **Asia**. In *Climate Change 2007: Impacts, Adaptation and Vulnerability. Contribution of Working Group II to the Fourth Assessment Report of the Intergovernmental Panel on Climate Change*, pp. 469–506, edited by M L Parry, O F Canziani, J P Palutikof, P J van der Linden, and C E Hanson. Cambridge, UK: Cambridge University Press

CWC (Central Water Commission). 2005. **Water Sector at a Glance.** Details available at http://cwc.gov.in/main/downloads/WaterData CompleteBook2005.pdf>, last accessed on 21 February 2010.

EEA (European Environment Agency). 2007. ***Climate Change: the cost of inaction and the cost of adaptation***. EEA Technical Report No. 13/2007. ISSN 1725-2237

Eijegeraam C J J J. 2005a. ***Cost-Benefit Analysis Room for the River, Part 2: cost effectiveness of measures and packages in Dutch***. Netherlands Bureau of Economy Policy Analysis. [CPB Document no 82, April 2005]

Eijegeraam C J J J. 2005b. ***Optimal Safety Standards for Dike-ring Areas***. Netherlands Bureau of Economy Policy Analysis. [CPB Document no. 62, March 2006]

Garg N K and Hassan Q. 2007. **Alarming scarcity of water in India.** *Current Science* **93**(7): 932–941

Gosain A K, Rao S, and Basuray D. 2006. **Climate change impact assessment of hydrology of Indian rivers.** *Current Science* **90**(3): 346–353

Central Ground Water Board. 1995. ***Ground Water Resources of India.*** New Delhi: Ministry of Water Resources

IPCC (Intergovernmental Panel on Climate Change). 1998. ***Regional Impacts of Climate Change.*** Cambridge: Cambridge University Press

IWMI (International Water Management Institute). 2009. **Strategic analyses of the national river linking project (NRLP) of India,** Series 1, In *India's Water Future: scenarios and issues*, edited by U Amarasinghe, T Shah, and R P S Malik. Colombo, Sri Lanka: International Water Management Institute. 417 pp.

Kirshen P. 2007. **Adaptation options and costs in water supply.** Details available at <http://unfccc.int/cooperation_and_support/financial_mechanism/financial_mechanism_gef/items/4054.php>. [Report to the United Nations Framework Convention on Climate Change (UNFCCC) Financial and Technical Support Division]

Kirshen P, Knee K, and Ruth M. 2008. **Climate change and coastal flooding in Metro Boston: impacts and adaptation strategies.** *Climatic Change* **90**(4): 453–447

Kríz B, Benes C, Cástková J, Helcl J. 1998. **Monitorováni epidemiologické situace v zaplaven´ych oblastech v Ceské Republice v roce 1997**. In Konference DDD '98; Kongresové Centrum Lázenská Kolonáda Podebrady, pp. 19–34, edited by P Davidová and V Rupes .

Kuik O. 2007. **Room for the River: a case study on the costs of adaptation to climate change in the Netherlands**. In *Costs of Adaptation to Climate Change: a review of assessment studies with a focus on methodologies used*. European Environmental Agency (EEA). [Agreement no 3602/B2005, EEA under the framework contract no EEA/AIR/04/004, Final working paper]

Kundzewicz A W, Mata L J, Arnell N W, Döll P, Kabat P, Jiménez B, Miller K A, Oki T, Sen Z, Shiklomanov I A. 2007. **Freshwater resources and their management.** In *Climate Change 2007: Impacts, Adaptation and Vulnerability. Contribution of Working Group II to the Fourth Assessment Report of the Intergovernmental Panel on Climate Change,* pp. 173–210, edited by M L Parry, O F Ganziani, J P Palutikof, P J van der Linden, and C E Hanson. Cambridge, UK: Cambridge University Press

McMichael A J, Campbell-Lendrum D, Kovats S, Edwards S, Wilkinson P, Wilson T, Nicholls R, Hales S, Tanser F, LeSueur D, Schlesinger M, Andronova N. 2004. **Global climate change.** In *Comparative Quantification of Health Risks: global and regional burden of disease due to selected major risk factors*, pp. 1543–1649, edited by M Ezzati, A Lopez, A Rodgers, C Murray. Geneva: World Health Organization

NATCOM. 2004. ***India's Initial National Communication to the United Nations Convention on Climate Change.*** Ministry of Environment and Forests, Government of India

Oxfam. 2007. **Adapting to climate change: what's needed in poor countries and who should pay.** Details available at: <http://www.oxfam.org.uk/resources/policy/climate_change/downloads/bp104_adapting_to_climate_change.pdf>, last accessed on 13 February 2010. [Oxfam briefing paper 104].

Parry M, Arnell N, Berry P, Dodman D, Frankhauser S, Hope C, Kovats S, Nicholls R, Satterhwaite D, Tiffin R, Wheeler T (eds). 2009. ***Assessing the Costs of Adaptation to Climate Change: a review of the UNFCCC and other recent estimates.*** London: International Institute for Environment and Development and Grantham Institute for Climate Change

Planning Commission. 2006. **Drinking water.** Details available at <http://www.nih.ernet.in/nih_rbis/india_information/drinking.htm>, last accessed on 23rd February 2010.

Stern N, Peters S, Bakhshi V, Bowen A, Cameron C, Catovsky S, Crane D, Cruickshank S, Dietz S, Edmonon N, Garbett S L, Hamid L, Hoffman G, Ingram D, Hones B, Patmore N, Radcliffe H, Sathiyarajah R, Stock N, Taylor C, Vernon T, Wanjie H, Zenghelis D. 2006. ***Stern Review: the economics of climate change.*** London: HM Treasury. Details available at <http://www.hmtreasury.gov.uk/independent_reviews/stern_review_economics_climate_change/sternreview_report.cfm>

UNDP (United Nations Development Programme). 2007. ***Fighting Climate Change: human solidarity in a divided world.*** New York: Palgrave Macmillan. 384 pp.

van Ierland E C, de Bruin K, Dellink R B, Ruijs A. 2006. **A qualitative assessment of climate adaptation options and some estimates of adaptation costs.** [ARK Netherlands Policy Programme, Wageningen]

Ward P J, Strzepek K M, Pauw W P, Brander L M, Aerts J C J H. In press. ***Costs of Adaptation Related to Industrial and Municipal Water Supply and Riverine Flood Protection.*** Washington, DC: The World Bank

World Bank. 2006. ***Investment Framework for Clean Energy and Development.*** Washington, DC: World Bank. 146 pp. Details available at <http://siteresources.worldbank.org/DEVCOMMINT/Documentation/20890696/DC2006-0002(E)-CleanEnergy.pdf>, last accessed on 4 March 2010.

World Bank. 2009. ***The Costs to Developing Countries of Adapting to Climate Change. New Methods and Estimates. The Global Report on the Economics of Adaptation to Climate Change Study.*** Washington, DC: The World Bank. 109 pp. [Consultation Draft]. Details available at <http://siteresources.worldbank.org/INTCC/Resources/EACCReport0928Final.pdf>

seven

Conclusion

Arabinda Mishra[a] *and Anil Markandya*[b]

Costing adaptation for key sectors is a prerequisite towards planning efforts on the national level for climate-resilient development. Since the projected outlays are huge in scale, financing opportunities and absorptive capabilities have become important considerations for developed and developing countries. For the Organisation for European Economic Co-operation (OECD) countries, linking adaptation to development has necessitated revision in the focus and direction of traditional development assistance for poverty alleviation and capacity creation. However, there continues to be considerable uncertainty over the choice of appropriate adaptation measures, owing to the lack of reliable climate change projections and limitations inherent to existing methods for vulnerability assessment. As one moves from the national to local levels of governance, there is a cascading effect with respect to the underlying uncertainty in sectoral impacts, assessments and, consequently, in cost estimates of adaptation.

The short-term approach to decision-making under uncertainty would be to go for "no-regrets" measures that have significant development side-benefits while contributing to greater resilience in the sectors. But, even for these measures to be effective, it may be necessary for existing policies and programmes to be reoriented and

[a] Director, Earth Science and Climate Change Division, The Energy and Resources Institute (TERI), and Dean, Faculty of Policy and Planning, TERI University.

[b] Professor, Ikerbasque, and Director, Centre for Climate Change (BC3), Bilbao, Spain.

create a consensus among the concerned stakeholders. It is difficult to estimate the transaction costs that would arise in this process of policy change, though a reasonable hypothesis would be that the more cross-cutting an impact is across sectors, the greater is the magnitude of such costs.

In relating development with adaptation, there may be an additional source of uncertainty to the kind of cost estimates presented in this study, which do not consider the possibility of non-uniform variations in the coping capacity of climate-risk-exposed communities over time. Economic development has a positive influence on coping capacity through technological progress, better access to resources, and stronger institutions. Except the agriculture sector, none of the other four sectors presented in this study have been examined for their potential to generate private autonomous adaptation responses, which are expected to be stronger and more wide-ranging with improved coping capacity. If autonomous and planned adaptation are linked to some extent—a possibility being increased autonomous responses resulting in lesser scope for planned interventions—cost estimates over a period of time are likely to show non-linear trends for different sectors.

Public funding for impacts and adaptation research is an immediate and cross-cutting need for all sectors in the country. It is, however, difficult to come up with a cost estimate for such research because of the methodological challenge of attributing specific outcomes to the scale and intensity of research effort. A typical approach is to use a normative criterion such as a certain minimum percentage of the country's gross domestic product (GDP), but this approach is inherently unsatisfactory.

Cost estimates of the kind presented in this study are expected to serve two primary purposes: (1) they inform the policy-makers of the net and additional investment required for adaptation measures in specific sectors for a given time period and (2) they force a discussion on financing issues. The Government of India currently projects about 2% of the country's GDP being earmarked for adaptation processes. This estimate is based entirely on public funding for *existing* programmes, whereas the need in the context of future climate change impacts is for new and additional financial resources. In international negotiations, on the issue of adaptation funding, India has made it clear that commitments from developed nations have to go beyond Overseas Development Assistance (ODA), and these should also be independent of market mechanisms. Presently, external funding for adaptation in India is available from a variety of bilateral and multilateral sources (Table 1), but our cost estimates suggest that it is still inadequate.

The public finance aspect of adaptation would call for an evaluation of existing budget allocations among sectors and the role of fiscal instruments in mobilizing additional resources, including some that do not burden the public finances. Mainstreaming adaptation action through government budgets is a practical requirement but would require policy reforms and enabling institutional arrangements at several levels of governance. An example in this context would be the phasing out of subsidies, particularly those that have led to adverse environmental and social outcomes in sectors such as agriculture and water. Resources freed up from such subsidy phase-outs could go towards funding adaptation measures. Insurance and micro-credit are important financing mechanisms that deserve greater attention from researchers concerned with mainstreaming adaptation.

The action on adaptation in India is largely going to be at the state level. There is, therefore, an urgent need for economic analyses of adaptation measures at the sub-national level. In principle, it should be possible to generate state-specific adaptation cost curves as decision support tools. In practice, a major constraint at present is the general lack of downscaled information on climate-related vulnerabilities. The climate modelling community has recognized this policy need, and efforts are being taken to come up with climate information for the future at finer spatial resolutions. The trade-off, as pointed out earlier, is in terms of a higher degree of uncertainty owing to inadequacies in the models and quality of input data. While acting on adaptation, the uncertainty factor needs to be appropriately weighted among options. Given that the economic analysis of adaptation is poorly understood in policy circles of India, there is a need for creating capacity at all levels of governance for effective decision-making in the context of future climate change.

There is considerable scope for refining both the methodology and estimates presented in this study. The value addition has been in terms of the key lessons derived from the research—(1) that the *scope* for government intervention on adaptation is large even in the

Table 1 Funding for adaptation: sources for India

Sl No	Source	Nature of disbursement	Sector focus	Availability criteria	Amount
National					
a.	Budgetary: central and state				Approximately 2% GDP to adaptation related activities
UNFCCC and KP					
a.	Adaptation Fund	Grants	Science, assessment, monitoring and early warning; tourism; water; terrestrial ecosystems; population and human settlements; disaster risk reduction; education and training; food security, agriculture, forestry and fisheries; health; oceans and coastal areas	Active civil society engagement with developing countries with regard to the call for proposals; carbon market	2% levy on clean development mechanism (CDM) proceeds
Multilateral					
1	Asian Development Bank (ADB)				
a.	Small Grants for Adaptation Actions (SGAs) in Asia and the Pacific	Grants		Grants available to ADB regional departments, national organizations, NGOs, private organizations, and academic institutions to conduct adaptation activities in developing member countries.	$1.2 million
b.	Climate Change Fund (CCF)		Fund available for technical assistance, investment projects, research and other activities		$40 million
c.	Water Financing Partnership Facility (WFPF)	Loans; technical assistance	Water		$65 million including adaptation

Continued...

SI No	Source	Nature of disbursement	Sector focus	Availability criteria	Amount
d.	Cooperation Fund for the Water Sector (CFWS)	Grants; technical assistance	Water		
e.	Poverty and Environment Fund (PEF)				$3.6 million, including adaptation
2.	Strategic Climate Fund (SCF) under Climate Investment Fund (Channelled by the ADB and World Bank Group)	Grants; loans	Terrestrial ecosystems; disaster risk reduction; education and training; food security, agriculture, forestry and fisheries; health; oceans and coastal areas;	Complement other multilateral financial mechanisms, such as GEF and Adaptation Fund. Overarching fund for various programmes to test innovative approaches to climate change. The first such programme is the Pilot Programme for Climate Resilience (PPCR)	India not yet a member
3.	The World Bank				
a.	Development Market Place (DM)	Grants	Disaster risk reduction Education and training		
b.	Global Facility for Disaster Reduction and Recovery (GFDRR)	Grants	Terrestrial ecosystems; disaster risk reduction; education and training; food security, agriculture, forestry and fisheries; health; oceans and coastal areas; population and human settlements; science, assessment, monitoring and early warning; tourism; water		

Continued...

Table 1 *Contd...*

Sl No	Source	Nature of disbursement	Sector focus	Availability criteria	Amount
c.	International Development Association (IDA)	Grants; loans	Terrestrial ecosystems; disaster risk reduction; education and training; food security, agriculture, forestry and fisheries; health; oceans and coastal areas; population and human settlements; science, assessment, monitoring and early warning; tourism; water		
d.	Program on Forests (PROFOR)	Grants	Food security, agriculture, forestry and fisheries		
4	Global Environment Facility (GEF)				Each GEF country member has a GEF operational Focal Point to ensure GEF projects are aligned to meet country needs and priorities.
a.	Special Climate Change Fund (SCCF)	Grants	Food security, agriculture, forestry and fisheries; terrestrial ecosystems; oceans and coastal areas; health	A development focused fund concerned primarily with long-term activities, programmes and measures in the development sectors that are most affected by global climate change	Financed by additional voluntary contributions channelled through GEF
b.	Strategic Priority on Adaptation (SPA)		Purpose is to ensure that climate change concerns are incorporated in the management o f ecosystems through GEF focal area projects	Financed by the GEF Trust Fund	
C.	Small Grants Programme (SGP)	Grants	Terrestrial ecosystems; water; population and human settlements; food security, agriculture, forestry and fisheries		

Continued...

Sl No	Source	Nature of disbursement	Sector focus	Availability criteria	Amount
Bilateral					
a.	Cool Earth Partnership	Grants; technical assistance	Food security, agriculture, forestry and fisheries; water; population and human settlements		Funded by Government of Japan
b.	MDG Achievement Fund (MDG-F)	Grants	Food security, agriculture, forestry and fisheries; water; terrestrial ecosystems; disaster risk reduction; health; oceans and coastal areas; population and human settlements; science, assessment, monitoring and early warning; education and training; tourism		Funded by Spanish Government; implemented by UNDP, UNICEF and FAO
c.	International Climate Initiative	Grants	Water; food security, agriculture, forestry and fisheries; disaster risk reduction; health; population and human settlements; terrestrial ecosystems		Funded by part of the auctioning revenues from emissions trading through the Federal Environment Ministry in Germany

UNFCCC – United Natiions Framework Convention on Climate Change; KP – Kyoto Protocol; GDP – gross domestic product; GEF – Global Environment Facility; UNDP – United Nations Development Programme; UNICEF – United Nations International Children's Emergency Fund; FAO – Food and Agriculture Organization; MDG – Millennium Development Goal short- and medium-term, with all its information constraints, (2) that the *nature* of such intervention would be additional to and different from business-as-usual development programming, and (3) that the *scale* of action required is large and likely to increase as we receive more knowledge about impacts and include more sectors for intervention (though development may act as a mitigating factor by building up coping capacity over time). The concept of adaptation is inherently dynamic in nature, and there would be a constant need to monitor and evaluate options and resource requirements, as the nature of vulnerabilities changes over time.

Index

Errata

Human Health

Page 65

Title of Annexure 3 should be read as

Some national programmes addressing issues in diarrhoeal prevention and treatment in India

Forest Ecosystem

Page 95

The equations to be read as

$$E_a = \frac{(I_1 - I_2)}{(I_1 - I_0)}$$

When the adaptation option completely avoids the impact of climate change (I_1), the effectiveness is equal to 1 and the expected impact with adaptation is equal to baseline.

$$\frac{(I_1 - I_2)}{(I_1 - I_0)} = 1;\ I_2 = I_0$$

Agriculture

Pages 131 and 132

The source does not pertain to Table 1

Coastal Zones

Page 184

Figure 2, page 184, legend

Port/Commb hub to be read as Port/Commercial hub